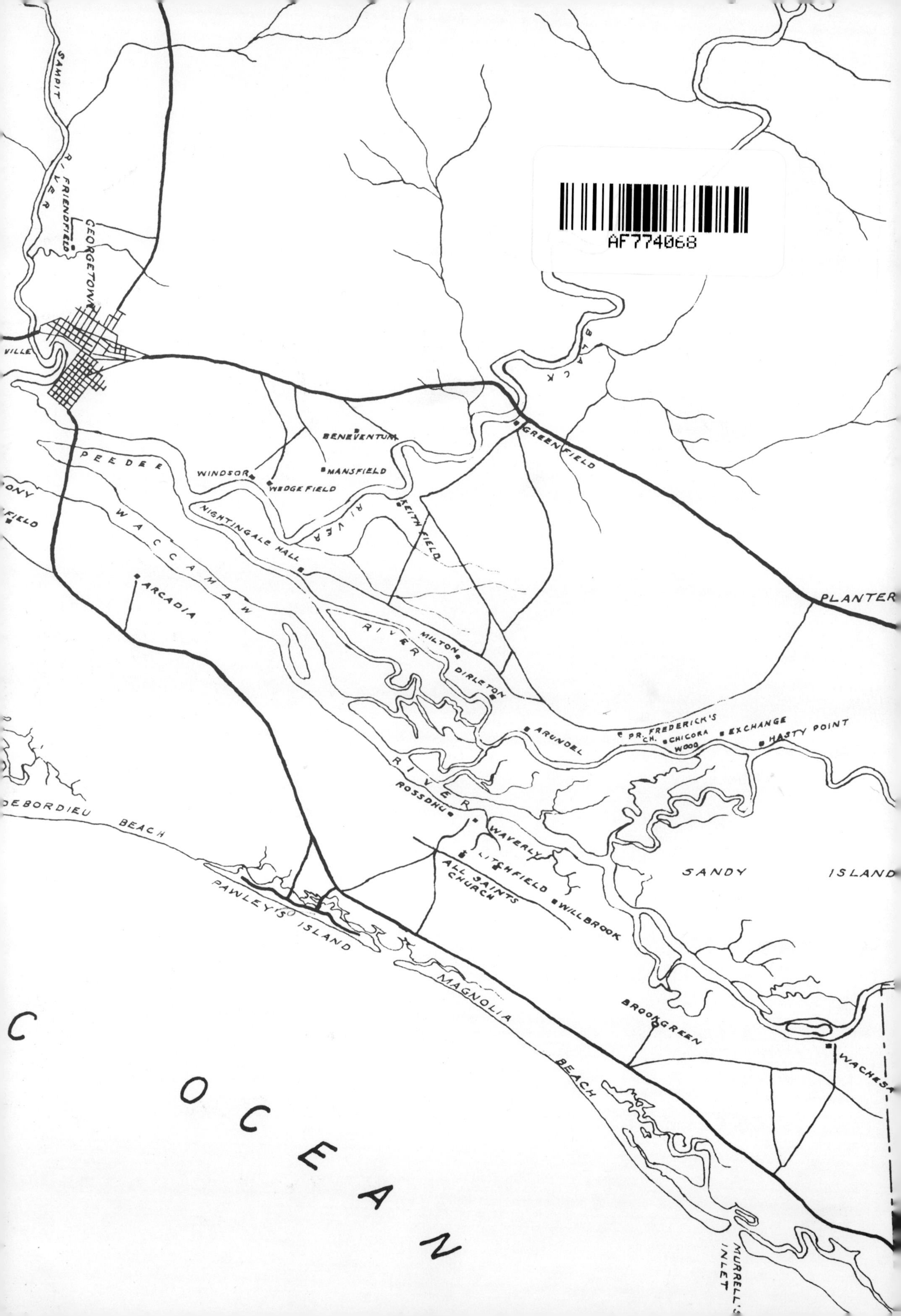
SAMPIT RIVER
FRIENDFIELD
GEORGETOWN
BLACK
BENEVENTUM
GREENFIELD
PEEDEE RIVER
WINDSOR
WEDGEFIELD
MANSFIELD
KEITHFIELD
NIGHTINGALE HALL
WACCAMAW RIVER
ARCADIA
PLANTER
MILTON
DIRLETON
ARUNDEL
PR. FREDERICK'S CH.
CHICORA WOOD
EXCHANGE
HASTY POINT
ROSSDHU
WAVERLY
LITCHFIELD
ALL SAINTS CHURCH
WILLBROOK
SANDY ISLAND
BEACH
PAWLEY'S ISLAND
MAGNOLIA BEACH
BROOKGREEN
MURRELL'S INLET
C
OCEAN

Alberta Morel Lachicotte

To

MY FATHER AND MOTHER

Arthur Herbert and Virginia Ingram Wilson Lachicotte

whose encouragement prompted

the writing of

Georgetown Rice Plantations

OLD RICE MILL AT CHICORA WOOD

GEORGETOWN

RICE PLANTATIONS

TEXT AND PHOTOGRAPHS
BY
ALBERTA MOREL LACHICOTTE

GEORGETOWN COUNTY HISTORICAL SOCIETY
GEORGETOWN, SOUTH CAROLINA
1993

© Alberta Morel Lachicotte 1955

First Printing, December, 1955
Second Printing, February, 1956
Third Printing [with revisions], December, 1961
Fourth Printing [with revisions], December, 1967
Fifth Printing, June, 1970
The State Printing Company, Columbia, S.C.
Charles E. Lee, Production Editor
Sixth Printing [with revisions],1989
The Reprint Company, Spartanburg, S.C.

Seventh Printing (with revisions), 1993
Georgetown County Historical Society
Georgetown, South Carolina 29442

New material copyright © Alberta Morel Lachicotte, 1993

Library of Congress Catalog Card Number 93-072943
Manufactured in the United States of America

Acknowledgments

THERE ARE MANY PEOPLE WHO ASSISTED IN VARIOUS ways toward the making of this book. For their graciousness and generosity, the writer is deeply grateful. First and foremost among these are the plantation owners who permitted their estates to be included and who cooperated fully in the search for material.

She is grateful also for the interest and assistance of Mr. Charles E. Lee, editor of the University of South Carolina Press, who edited the manuscript and supervised publication.

Officers of the Woman's Auxiliary of Prince George, Winyah, Protestant Episcopal Church, sponsoring group of the annual Spring Plantation Tours, encouraged the writing of this book, and it was through work with them that the writer became interested in Georgetown's plantations. She therefore is indebted heavily to that organization.

For her generous spirit, encouragement and advice, Miss Katharine M. Jones of Greenville is extended a special measure of gratitude. The author wishes to express her appreciation also to Mr. Herbert Ravenel Sass of Charleston and to Dr. J. H. Easterby, Mrs. Arney R. Childs, and Dr. Chapman J. Milling of Columbia for certain important information they very graciously contributed.

She would like to thank, too, Mr. T. Cordes Lucas of Georgetown County for his invaluable assistance with the Winyah Bay and North Santee plantations and Mr. James R. Parker of Georgetown for his help in gathering data on the Black River properties. Miss Susan Lowndes Allston of Waccamaw and of Flat Rock, N.C., was of great assistance in contributing data on the Allston family and on various plantations. The writer is indebted not only for her help but also for her deep interest.

Mrs. Frederick Wentworth Ford and the late Mrs. Charles Weston Rosa of Georgetown, Miss Emma B. Richardson of Charleston, Mr. Robert V. Read of Columbia, the Rev. Dr. H. D. Bull of Barnwell, the Rt. Rev. A. S. Thomas, retired Bishop of the Episcopal Diocese of

South Carolina, and Mr. E. Milby Burton, director of the Charleston Museum, all gave readily and fully from historical data available to them.

Others to whom the writer is indebted for material, for assistance in locating material, or for help in other ways are Captain and Mrs. Joshua John Ward, Mrs. John E. Allston, Mr. Richard L. Allston, Miss Florence Lachicotte, Mrs. Russell E. Davis, and Mr. Arthur LaBruce, all of Pawleys Island; Miss Margaret B. Rouquie, Miss Minnie Hazard, Dr. F. A. Bell, the Rev. A. N. Daunt, Mrs. Harold Kaminski, Mr. Nat Kaminski, Jr., Mrs. Tom Skinner, Mrs. H. D. Oswald, Miss Harriet Plowden, and Mr. and Mrs. Robert H. Ellis, all of Georgetown; Mr. T. N. Cox, superintendent of Arcadia plantation; Mr. Patrick C. McClary, Jr., manager of Friendfield plantation; Mr. D. L. Vaill, Jr., secretary at Dover; Mr. Frank G. Tarbox and Mrs. Genevieve W. Chandler of Brookgreen Gardens; Mr. Francis Johnstone of Athens, Ga.; Mrs. Horace B. Butterfield of Tacoma, Wash.; Mrs. L. B. Middleton of Charleston; Mr. Frank Vance of Columbia; Miss Genevieve Chandler, Mrs. Robert S. Bonds, and Mrs. Effie Cannon Bruns Thatcher, members of the staff of the Georgetown County Memorial Library; the staff of the clerk of court's office at the Georgetown County Courthouse; and the staff at the South Carolina Historical Society library in Charleston.

The writer's debt to the researches of her predecessors in the study of Georgetown County history is demonstrated by the notes at the back of the book. The University of Chicago Press granted permission to quote material from J. H. Easterby's *The South Carolina Rice Plantation: As Revealed in the papers of Robert F. W. Allston* (copyright 1945 by The University of Chicago) and from Edgar P. Richardson's *Washington Allston: A Study of the Romantic Artist in America* (copyright 1948 by The University of Chicago); the University of South Carolina Press, from Arney R. Childs' *Rice Planter and Sportsman: The Recollections of J. Motte Alston, 1821-1909* (Columbia, 1953); Carter Field, from his *Bernard Baruch: Park Bench Statesman* (New York: McGraw-Hill Book Co., Inc., 1944); Miss Susan Lowndes Allston, from her *Brookgreen, Waccamaw, in the Carolina Low Country* (1935) and from Elizabeth W. Allston Pringle's *Chronicles of Chicora Wood;* the Rev. Dr. Henry D. Bull, from his *All Saints Church, Waccamaw, 1739-1948* (Published by the Historical Activities Committee of the South Carolina Society of Colonial Dames of America, 1948) Mrs. John E. Allston, from her *Allstons and Alstons of Waccamaw* (1936); and Brookgreen Gardens, from its published pamphlets.

Contents

PEEDEE RIVER (continued)

SAMPIT RIVER AND WINYAH BAY

NORTH SANTEE

Illustrations

GEORGETOWN RICE PLANTATIONS

A RICE FIELD AT WAVERLY PLANTATION (CA. 1890)—
TYPICAL OF MOST LOWCOUNTRY RICE PLANTATIONS

Introduction

MANY YEARS HAVE PASSED SINCE THE RIVERS OF GEORGE-town County echoed with the voices of oarsmen, singing as they poled their rice barges up and down the broad streams and through the numerous creeks that wound across the vast rice fields along the shore. Yet that bygone age of rice planting and great mansions still lives—in the memories of a few, in the romantic hearts of the present generation, and in the land itself, which seems to recall longingly its lost mode of life.

For decade after decade, from the last years of the eighteenth century through the opening of the twentieth, there flourished in this county of South Carolina the most productive rice culture of the South Atlantic coast, and one of the most productive in the whole United States. From the very early years of settlement until the Revolution, indigo, a blue dye, was a major money crop. Then, competing dyes and the lost English bounty made it unprofitable, and rice, which already had a foothold, took over completely and reigned supreme.

The terrain was ideally suited for the development of a rice industry. Four rivers, the Waccamaw, the Peedee, the Black, and the Sampit, brought their waters down the length and breadth of the county, converging at the town of Georgetown to form Winyah Bay; and only a few miles to the south the great delta of the Santee provided seemingly unlimited space for the growing of the profitable grain. On these low, muddy lands fortunes could be made; and along the shores of the rivers rice plantations sprang up, one after another, as far as the tides could go. As wealth increased, the planters built fine houses, surrounded by landscaped gardens and approached by live-oak avenues, and provided themselves with all the comforts of gracious living. A way of life, as well as an economy, based on rice came into existence.

On the whole, it was a stable life and economy. True, fortunes could be made one year and lost the next by an untimely freshet, which caused often the mortgaging and occasionally the loss of family estates. Yet rice planting survived these minor disasters and the major one of the War Between the States and continued to dominate the region throughout the remainder of the century. Then, abruptly, in 1908, planting almost ceased. Three years of severe freshets and hurricanes, beginning in 1906, left rice fields in ruins, crops utterly destroyed, and planters bankrupt. These three seasons of devastating weather are considered locally more responsible for the death of the rice industry in South Carolina than the often blamed competition of Texas, Arkansas, and Louisiana, where the grain was being grown more cheaply with the use of machinery. Struggles to revive planting in Georgetown County continued through 1911. Then rice, so long the mighty king of the low country, disappeared, leaving no heir to take its place.

The years have shrouded the plantation era in mystery and romance; and old ruins and overgrown gardens only deepen the appeal of imaginary tales so easily inspired by the historic countryside. Because these plantations are so cloaked in legend, it is often difficult to discern where fact stops and fiction begins; and though the old estates are the more interesting for those embellishments, it is not the purpose of this book to set down in print these stories of romance: they will be with us always and with the passing of time will grow even more intriguing.

Rather, the writer has tried to record in permanent and accessible form the verified histories of the plantations so that the authentic may not be lost entirely in the fictional. In her endeavor to be as accurate as possible, she has consulted the most reliable sources available. Old deeds, wills, and other family documents have been drawn upon heavily. Official Georgetown District records for the period between 1785 and 1865 were destroyed during the War Between the States, when they were removed inland for safe-keeping and where they promptly fell into the path of invading Union armies; so private sources of information have been relied upon to a great extent to fill the vacuum. Strenuous efforts have been made to check the accuracy of dates, but research revealed occasional discrepancies even in church records, proving that human imperfection is not a monopoly of the present era. The writer can but hope, therefore, that errors have been reduced to the lowest possible minimum.

This book attempts to consider only the rice plantations of Georgetown County that still exist and the properties they have absorbed from others that no longer possess dwellings. Many estates, having lost their residences either through decay or fire, are now uninhabited wildernesses. It would be a task of years and of questionable value to relate the histories of those places in addition to the presently existing plantations.

Tracing the ownerships and acreages of Georgetown plantations has been a pleasant task but an arduous one: many estates went through a complicated history of division and subdivision and of union with other estates that both fascinates and exasperates the researcher. Because of these vicissitudes of land-holding and because a number of plantations often grew out of a single original grant, it is often impossible to tell the story of one plantation without repeating part of the story of another. It is hoped that these repetitions will be less irritating to the reader than the numerous crossreferences that would have been otherwise necessary.

The period of rice planting was the age of magnificence in the low country. There probably never again will be in this region a culture to equal in glamor plantation life. It is the writer's hope that these pages will contribute toward a truer conception of that great era as it existed so long on the upper coast of South Carolina.

Introduction to Sixth Printing, 1989

Change—in varying degrees, but change nevertheless—is the aspect above all others that comes to mind when one considers today the old rice plantations of Georgetown County. One of the greatest rice-producing areas in the United States prior to the Civil War, Georgetown County also boasted some of the richest planters in the nation. From the early seventeen hundreds until the end of rice planting in 1911, plantations lined the banks of the county's waterways—the Waccamaw, Peedee, Black, Sampit and North Santee rivers and Winyah Bay. These plantations, ranging from small to very large, earned their masters enormous wealth, wealth that built impressive homes and financed luxurious lifestyles. From 1911 until the end of World War II, however, there was no economy to take the place of rice; many families lost their plantations through mortgage foreclosures. The once-flourishing country estates just sat, for the most part, falling gradually into decay. Most were used during this time for hunting.

The postwar period of the 1940's gave rise to a new era, the vacation resort economy. The summer resort business had actually begun in the 1920's and 1930's along the beaches from Pawleys Island northward. Development continued during the late 1940's and the 1950's. Over the past twenty-five years, however, the market for waterfront property and resort housing—with the accompanying facilities of golf courses, tennis courts, club houses and marinas—has gained astonishing momentum. Former rice plantations are being transformed right and left into exclusive resort residential communities. Nowhere is this kind of change more the case than on Waccamaw Neck, the long, three-mile-wide peninsula between the Waccamaw River and the Atlantic Ocean and stretching from the mouth of Winyah Bay to Georgetown County's northern boundary. Brookgreen Gardens (formerly Brookgreen plantation) and Hobcaw Barony (controlled by the Belle W. Baruch Foundation) stand now as the two bastions of stability amidst an onslaught of developments. There are already eight golf courses on Waccamaw Neck, and two more are scheduled. Surrounding each, naturally, is a residential community.

The major factor in the rapid growth on Waccamaw Neck is the river's proximity to the ocean. The whole peninsula is part of the resort area. The Black River and Winyah Bay plantations close to Georgetown are the places next most affected by change. These are becoming residential subdivisions that are really extensions of the town. The least-changed plantations in the county at the present time are those on the Peedee, the North Santee and lower Winyah Bay. Their distance from both the beaches and Georgetown probably accounts for their lack of change. The bastion of stability on lower Winyah Bay and stretching toward the Santee is the Tom Yawkey Wildlife Center. It includes North and South Islands and most of Cat Island, all three situated at the mouth of Winyah Bay. These holdings encompass some 20,000 acres. Brookgreen, Hobcaw Barony and the islands of the Yawkey Wildlife Center are supposedly protected from further change. The Peedee and North Santee areas are not; these plantations may some day be affected like the others.

It goes without saying that the local growth of the last quarter-century has brought great prosperity, in fact, more widespread prosperity than ever before. The price has been the significant loss of plantation charm and picturesque beauty, the qualities that have made the plantation sections of Georgetown County so appealing to the people who come here to see them. Such is the ultimate irony! Yet, these areas cannot have life both ways.

Because of the change at so many plantations since the first printing of *Georgetown Rice Plantations* in 1955 and especially since the fifth printing in 1970, this, the sixth printing, is a mixture of past and present. If the book were to be updated in every way, nothing would be left at some plantations but residential communities, clubhouses, golf courses or a combination thereof. I have tried to do with each place what seemed to be the reasonable, manageable thing from the standpoint of organizing new material. Future changes may present even greater complications for later printings of this book. For now, both author and reader are fortunate that things are as understandable as they are. This printing offers an updating of the plantations from June of 1970 through December of 1988.

Alberta Lachicotte Quattlebaum
Waverly Plantation
December 31, 1988

Editor's Note

This edition of *Georgetown Rice Plantations* has a new introduction to the Sixth Printing and 32 pages of addenda which update the information in the Fifth printing of 1970. The addenda are placed at the end of each section—Waccamaw Neck Addenda, pp. 68a–68k; Black River Addenda, pp. 96a–96d; Peedee River Addenda, pp. 132a–132e; Sampit River and Winyah Bay Addenda, pp. 158a–158c; and North Santee Addenda, pp. 194a–194i. Certain of the photographs in this volume are also new; this will be indicated in a note beneath the picture. A new supplementary index has been added and has been combined with the supplementary index which was prepared for the third printing and which has appeared in previous editions; the consolidated supplementary index is placed at the end of the regular index, beginning on p. 222.

WACCAMAW NECK

Hobcaw Barony

THE VAST ESTATE OF HOBCAW BARONY, STRETCHING FOR miles down the Waccamaw peninsula and from Winyah Bay to the sea, is well known as the plantation of Bernard M. Baruch, America's Elder Statesman and Adviser to Presidents. Baruch, born in Camden, South Carolina, has associated with many international figures and with all the American Presidents from Woodrow Wilson to Dwight D. Eisenhower. On two occasions between the world wars he was host at Hobcaw to the great British statesman Sir Winston Churchill; and President Franklin D. Roosevelt shortly before his death stayed a month at the plantation and on various outings visited many other places on Waccamaw Neck.[1]

The illustrious present of the barony is matched by its equally illustrious beginning. One of ten baronies of 12,000 acres each, laid out in 1711 and divided among the Lords Proprietors by lot, Hobcaw, called by its Indian name, fell to John, Lord Carteret, and was granted formally to him on December 5, 1718.[2] Carteret, later the Earl of Granville, has been described as "'one of the first orators, purest patriots, keenest wits, brightest classical scholars, and most ardent convivialists of his time.'"[3] In the years to come characteristics similar to his would be found in abundance among the inhabitants of Waccamaw Neck, but in his day the area was a wilderness and unsuitable for the exercise of his talents. Carteret never saw his American estate, and after holding it for twelve years, he sold it in 1730 to John Roberts of Dean's Court, Middlesex, England.[4]

On a resurvey applied for by Roberts in 1731, the tract was found to contain 13,970 acres, and Roberts received in 1736 a formal grant for the additional acreage. From Roberts, Hobcaw passed to Sir William Baker, Nicholas Linwood, and Brice Fisher—all three apparently London merchants. In 1765 these men appointed Francis Stuart and Paul Trapier, a Georgetown merchant, as their attorneys with full power to sell the property, but Stuart died soon afterwards, leaving Trapier to make all sales.[5]

HOBCAW BARONY

At this point there began a series of involved divisions and subdivisions. Trapier made his first sales from the barony in November, 1766, and by the last of January, 1767, had disposed of the entire 13,970 acres. The new owners, with their portions, from the northern boundary to the southern tip, were Thomas Mitchell, 2,412 acres; Robert Heriot, 2,177; Benjamin Huger, 1,711; Henry James Daubuz, an estimated 1,789¾; Peter Secare, 1,061½; Benjamin Trapier, 1,515; and Samuel Clegg, 3,303¾.[6] Hobcaw's later history makes necessary a discussion of each of these tracts.

The Mitchell Tract:—Thomas Mitchell died, it is believed, in 1768, leaving a will devising his tract to his son Edward, who divided the land and in March, 1785, sold the northern half, 1,206 acres, to William Alston, son of Joseph Allston of The Oaks plantation. The purchase, which became known as Clifton, stayed in the Alston family until after 1900,[7] when it was annexed to adjacent Prospect Hill along with other northernmost places developed from Hobcaw Barony to form Arcadia plantation.

The southern half of Edward Mitchell's tract, also 1,206 acres, was sold on February 4, 1786, to John Allston. Known thereafter as Forlorn Hope, the property seems to have passed to Allston's daughter Mary, who married Benjamin Huger, son of Major Benjamin Huger of Revolutionary fame. Some 793 acres of this tract, under the name Forlorn Hope, was conveyed in 1835 by the executors of Huger's estate to William Algernon Alston,[8] son of William Alston of Clifton. The plantation now is part of Arcadia.

The Heriot–Huger Tracts:—Robert Heriot on January 15, 1772, acquired the portion of Hobcaw originally sold to Major Benjamin Huger—making the Heriot holding 3,888 acres in all. After Robert's death the northern 1,243 acres went first to Roger Heriot and then to Mrs. Mary Heriot, who sold it on January 1, 1802, to Colonel William Alston of Clifton. This became Rose Hill, and a year to the day from his date of purchase Colonel Alston transferred the plantation to his son William Algernon Alston.[9] Rose Hill is also part of Arcadia today.

The next portion of the Heriot property to be sold became Alderley plantation. No date for the sale is on record, but the place was apparently in the ownership of Benjamin Huger II in 1802. By 1808 it had passed into the hands of Colonel Francis Kinloch Huger, youngest son of Major Benjamin Huger. Colonel Huger had distinguished himself in 1794 by an unsuccessful attempt to free the Marquis de

Lafayette from the Austrian prison fortress of Olmutz. He later served in the War of 1812.[10]

The part of the Heriot land remaining after these two sales was conveyed in two separate tracts to Thomas Young. The northernmost, called Annandale, following Young's death was resold in separate parcels—the portion on the river later known as Oryzantia being purchased by Benjamin Allston, and the pinelands and seashore area going to Colonel William Alston of Clifton. Youngville, the second tract bought by Young from the Heriot estate in 1794, was also acquired by Colonel Alston, who purchased the pinelands and seashore part directly from Young's estate and the river portion from Benjamin Allston, who owned it for a while after Young's death.[11]

The Daubuz–Secare Tracts:—One part, 870 acres, of Henry James Daubuz' tract became Bellefield. It too became the possession of Thomas Young and later, in 1808, of Colonel William Alston. The rest of the Daubuz land was apparently added to the Peter Secare tract, and the whole became Marietta plantation.[12]

The Trapier Tract:—Benjamin Trapier conveyed his purchase of the Hobcaw grant to William Burnett, who in 1784 divided the land, selling the upper part, 746½ acres, later called Friendfield, to Edward Martin, and the lower area of 768½ acres, afterwards named Strawberry Hill, to Peter Foissin.[13]

The Clegg Tract:—Samuel Clegg's portion of the barony also became divided. The western part called Calais, facing Winyah Bay, belonged in 1796 to the Rev. Hugh Fraser, rector of All Saints, Waccamaw, Protestant Episcopal Church. The area often has been called Fraser's Point. The eastern part of Clegg's property went to a man named Michaux.[14]

So it was that the grant of land on Waccamaw made out to Lord Carteret became divided and subdivided. In 1857 William Algernon Alston of Rose Hill acquired portions—or perhaps all at that time—of the lower plantations developed from the barony. These were Marietta, Strawberry Hill, Friendfield, Michaux's Point, and Calais. It is known definitely that he owned them by 1860, the year of his death. These plantations continued in the William Algernon Alston estate until 1875, when they were sold by his executors to the Scotchman Robert James Donaldson, along with other tracts on the barony.[15] His two sons, John Henry and Sidney T. Donaldson continued to plant rice on the estate until 1905, when they sold the estate to Bernard Baruch.[16]

Bellefield, Youngville, and Annandale apparently passed from Colonel William Alston to his son Charles, and in turn to Charles' son Charles Pringle Alston, who with his sister Susan Pringle Alston sold all three places with certain seashore tracts to Mr. Baruch in 1905.[17]

Baruch completed his acquisition of Hobcaw lands in 1907 with the purchase of Alderley from Edwin W. Kaminski. Colonel Francis Kinloch Huger sometime before 1848 had sold Alderley to Colonel Joshua John Ward of Brookgreen plantation. Ward also added to his holdings Oryzantia plantation, which had passed sometime earlier out of the hands of Benjamin Allston. (John Izard Middleton of Bannockburn plantation is known to have had possession of Oryzantia in 1838.[18]) Ward died in 1852, but his estate was not officially settled until 1867. The northern portion went to a daughter, Margaret Bentley Ward. The remainder had been willed to Colonel Ward's son Mayham but since Mayham died before the final settlement, the property went to his widow, Florence Lee Ward, and his son, Mayham, Jr. This son died young and unmarried, and Mrs. Ward's holding passed to her niece and namesake, Florence Ward. By 1905, the entire Alderley estate, including the northern portion, Oryzantia on the south, and the central portion (sometimes called Middle Square) had come into the possession of Samuel Mortimer Ward, Florence Ward's brother. Shortly afterwards, the properties passed to Edwin W. Kaminski, a prominent Georgetown businessman.[19]

With the Donaldson, Alston, and Kaminski purchases, Bernard Baruch came into ownership of all the original Hobcaw Barony, with the exceptions of three plantations, Clifton, Rose Hill, and Forlorn Hope.

For years Baruch lived in the residence on Friendfield built by Robert Donaldson and used by the Donaldson family during their ownership.[20] The house burned in 1930 and was replaced in 1931 with the present stately mansion. Mrs. Raoul Fleishman, who is now one of the owners of *The New Yorker* magazine, was an interior decorator in 1931, and she oversaw the finishing of the new house.[21] The gleaming white columns, which extend almost the entire width of the front, form with the beautifully landscaped grounds and superb natural setting of live oaks an impressive scene on the high bluffs of Winyah Bay. Located at the north edge of the Hobcaw gardens is a large doll house which Baruch built for his daughter Belle. Beyond the grounds is a tennis court used at one time by the family and guests.

Hobcaw has been honored with visits by many famous personages. Generals George C. Marshall and Omar Bradley have been among the guests, and while President Roosevelt visited at the barony in World War II, scores of high officials, scurried in and out on important missions.[22]

Innumerable Senators have been at Hobcaw on various occasions. An amusing incident occurred when one group came on a duckhunting trip during the early days of Prohibition.

Hucks Caines, one of three brothers—all colorful, veteran duck hunters and guides—had taken four Senators from very, very dry states out on a morning expedition. As the men were returning from the unsuccessful outing, Senator A. O. Stanley of Kentucky drifted into conversation with Caines. He introduced the other three Senators to the guide and told him that they were the men who made the laws up in Washington.

Grunting skeptically, Caines asked, " 'Kin I say sump'in?' "

" 'The sky is the limit,' " Stanley jovially replied.

" 'Well,' " said Caines, " 'If you all don't know no more about other things than you do 'bout ducks and whiskey, this country is in a hell of a fix.' "[23]

Some years earlier another brother, Sawney Caines, an equally irrepressible guide, made very frank remarks to an official higher than the four Senators. The man was no one less than President Grover Cleveland. In 1894 the Annandale Gun Club, composed of men from Philadelphia and New York, entertained Cleveland at the clubhouse on the Annandale portion of the barony. It was the President's first duck hunt, and Caines had been chosen as guide for his group.[24]

As the story is told, the President's party had been in the rice fields for some time on a cold December morning. Sawney first had to dislodge the small boat from a mud bank then he had been called upon to free the heavily-booted President from the same mud. These little annoyances Caines bore with manly restraint, but the one sin that he, an expert marksman, could not forgive was missing a duck. Cleveland, a notoriously poor shot, proceeded to do the unforgivable. He missed what was to Sawney an easy target. As the shots echoed loudly and vainly over the wide expanse of watery marsh, a feeling of pain overwhelmed the disgusted Caines. Unable to contain himself any longer, the guide blared out, "Sit down and gimme dat gun. Even if yo' is de Presi-dent of the Nunited States, yo' ain't worth a damn!" President Cleveland sat.

In the last several years, the great Hobcaw mansion has not been used so frequently as in former days. Instead, Baruch now spends most of each winter at a smaller place near Kingstree, South Carolina.

Bellefield

THE HISTORY OF BELLEFIELD PLANTATION IS TO A GREAT extent the same as that of Hobcaw Barony since Bellefield originally came from that large colonial grant and later was re-absorbed into the one holding. For some years Bellefield has been the winter home of Miss Belle W. Baruch, to whom it was conveyed by her father, Bernard M. Baruch.

The handsomely styled residence completed about 1939 by Miss Baruch, is situated in a large grove of live oaks. On one side, a spacious terrace banked with azaleas overlooks a wide shaded lawn, which slopes beneath towering oaks and pines to a lake set in the great forests bordering Winyah Bay. Located just beyond the grounds and gateway are stables where the plantation's fine saddle horses are housed. Neatly kept grounds, a blessed natural setting, and a stately dwelling make Bellefield one of Waccamaw's loveliest homes.

The land that developed into Bellefield goes back to the years 1766 and 1767, when the divisions and sales of Hobcaw Barony's 13,970 acres took place. Henry James Daubuz purchased 1,789 acres of the property and afterwards divided his acquisition into two parts. One portion, containing 870 acres went to form Bellefield. The other, combined with land from Hobcaw Barony originally sold to Peter Secare, became Marietta.

Thomas Young, who owned Annandale and Youngville, two other plantations developed from Hobcaw Barony, also at one time owned Bellefield. After his death, all three plantations eventually became the property of Colonel William Alston of Clifton. From him they apparently passed to his son Charles, who made his home at Bellefield[25] and left the place upon his death to his children, Charles Pringle Alston and Susan Pringle Alston, who sold Bellefield along with Annandale and Youngville to Bernard Baruch in 1905.[26]

Thus, the barony has been divided a second time. Miss Belle Baruch now owns the whole northern part, and her father the southern portion.[27]

BELLEFIELD, VIEWED FROM THE LAKE

Arcadia

TWO PLANTATIONS WHERE PRESIDENTS OF THE UNITED States were entertained, Clifton and Prospect Hill, now rest in impressive dignity within Arcadia's extensive boundaries. Both of these rice plantations entered history at an early date—at the very time that they were being groomed into low-country show places by the construction of fine mansions and the landscaping of gardens. George Washington stayed overnight at Clifton during his southern tour of 1791; and following him to Waccamaw in 1819, James Monroe was a guest at Prospect Hill, where his arrival and departure were staged with full pomp and circumstance.

The main seat of Arcadia, owned at present by Mrs. Philip H. Brady, Jr., is the original Prospect Hill mansion and gardens. From here the property fans out to include the former plantations of Clifton, Rose Hill, Forlorn Hope, George Hill, Fairfield, Oak Hill, and Bannockburn. It also reaches to the sea, taking in part of Debordieu Beach. During the ownership of Dr. Isaac E. Emerson of Baltimore from 1906 until 1931 the whole estate was given its present name of Arcadia.

So prominently connected with Arcadia is the Allston (or Alston) family that the story of the plantation would be incomplete without at least a brief excursion into Allston history, for it was under this family that the place had its beginning.

John and William, the first of the Allston male line to settle on Waccamaw Neck, acquired lands here through grants and purchases between 1730 and 1750. They were the sons of John Allston of St. John's Parish, Berkeley, who had come to America from England in the late 1600's. The brothers probably first settled on Waccamaw at The Oaks and Turkey Hill, as is indicated by the old family burial plots at these two plantations.[28]

William was married in 1721 to Esther LaBrosse de Marboeuf. (The LaBrosse name later became LaBruce.) The fifth of their many children, Joseph, born 1735, was the first to own Prospect Hill as a single tract. Known in the family as Joseph Allston of The Oaks, this son purchased tracts totaling about 1,400 acres from John Huger, Thomas Butler, and George Smith. To his eldest son, William, Joseph willed the northern half of this property, the portion afterwards called Fairfield. To his fifth child, Thomas (born in 1764), the only one besides William to live to maturity, Joseph willed the southern half, and this part became Prospect Hill. On the death of his father in 1784, Thomas came into possession of both Prospect Hill and Turkey Hill farther up the Waccamaw.[29]

Thomas married his cousin Mary Allston, daughter of his Uncle John, nicknamed during the Revolution, "Captain Jack of the Raccoon Guards." They established residence at Prospect Hill and built here the beautiful two-story mansion which stands today. It is a duplicate of the house which Thomas' brother William constructed at Clifton just to the south.

Thomas had no children, and at his death in 1794 the still unfinished house and the plantation, then about 550 acres, went to his widow. She later married Benjamin Huger II, son of Major Benjamin Huger, who had achieved fame as the American to greet the Marquis de Lafayette on the latter's arrival at North Island during the Revolution. On her second husband's death, a few years following the marriage, Mrs. Huger sold the plantation to Colonel Joshua John Ward of Brookgreen.[30]

The famous gardens at Arcadia were apparently started in Mrs. Huger's time, since the published reminiscences of W. Hasell Wilson; a Waccamaw visitor of that period, mention Prospect Hill, as the home of Mrs. Huger and say, "There was a handsome garden attached to the house, and it was considered quite a showplace."[31]

President Monroe's welcome to Prospect Hill on April 21, 1819, was an elaborate one. One story of the great event says that a carpet for Monroe to walk upon was laid from the front entrance of the house to the canal, the direction from which the President's party came. Possibly the carpet has been extended a few feet with each telling of the tale, for laying such a cloth even halfway to the canal two hundred yards away would have been task enough and suitably impressive. But the plantation gentry were fond of doing things in the grand manner, and the President may indeed have walked on carpet all the way from canal to mansion. The welcome of Monroe's party was equaled by the lavish departure to Georgetown next morning in "one of the plantation barges, profusely decorated and adorned for the

ARCADIA

occasion with the United States colors proudly floating at its head. Eight negro oarsmen dressed in livery propelled the barge."[32]

Mrs. Huger after selling Prospect Hill to Colonel Ward went to Charleston to live and died there in 1838, leaving no children. However, she left a legacy of $5,000 to the parish of All Saints, Waccamaw, making possible the razing of the wooden building then in use and the construction of a new, brick church. This handsome large church burned in 1915 and was replaced by the present smaller one built along similar lines.[33]

Colonel Ward in his will dated 1848 left Prospect Hill to his son Benjamin Huger.[34] Though the Colonel's death occurred in 1852, his heirs did not gain official possession of their various inheritances until after the War Between the States, when all had come of age. The Benjamin Ward family lived at the plantation and planted rice for some years. When Dr. Emerson purchased the property in 1906, following the long Ward ownership, the land consisted of 1,060 acres.[35]

Dr. Emerson in a few years added tremendous acreage to his holding through the purchase of six neighboring plantations. He made the house at Prospect Hill the main residence, enlarging it by the addition of wings on both sides. He also expanded the garden and named his part of the development the Arcadia gardens. The grounds at Arcadia are among the most beautifully landscaped in South Carolina. Terraces, pools, and gateways branch out from the portico and side wings and extend to the rice fields. Centering the third terrace is a star-shaped reflection pool. Huge azaleas bank each section of the gardens, and camellias line the pathways and drives. When the undergrowth on the north side of the house was being cleared away some years ago, there were found the banks and ditches of an old garden, believed to be part of the gardens started by Mrs. Huger, for the arrangement of banks seemed to form the letter H. Altogether the gardens contain thousands of plants, and more are added periodically from the plantation greenhouse and nursery maintained for the purpose. A professional nurseryman has the full-time task of supervising the care and further development of the grounds. Some authorities on landscaping place these gardens in a class with Magnolia and Middleton.

In 1906, the year of his purchase of Prospect Hill, Dr. Emerson also bought, from Doyle Land and Timber Company, Oak Hill plantation, 3,425 acres, part of the old John LaBruce estate.[36]

In 1909 he purchased four more plantations, Forlorn Hope, George Hill, Rose Hill, and Clifton. Forlorn Hope, containing 793 acres, was originally part of Hobcaw Barony. It was conveyed to William Algernon Alston in 1835 and sometime after his death in 1860 passed to three heirs, Margaret D. Lucas, Josephine A. Simons, and Mary A. Deas, who sold it to Emerson.[37] George Hill was obtained from Martha A. Pyatt, Penelope B. Parker, B. Allston Pyatt, Catherine W. Pyatt, and Maham W. Pyatt.[38] Rose Hill, a plantation of 1,243 acres, which also came out of Hobcaw Barony, became the home of William Algernon Alston upon transfer from his father, Colonel William Alston of Clifton. William Algernon built the residence there—a roomy, attractive home, which burned in 1906 during the visit of a hunting party.[39] Emerson purchased the property from the heirs of the William Algernon Alston estate—Helen Alston, Fanny Alston, Rowland Alston, Thomas L. Alston, and William A. Alston.[40]

The historic Clifton plantation was once the home of Colonel William Alston and played a large part in the story of this outstanding planter, the eldest son of Joseph Allston of The Oaks. William (born in 1756) inherited Fairfield upon his father's death in 1784. Instead of settling there, however, in 1785 he bought 1,206 acres of the northernmost part of Hobcaw Barony from Edward Mitchell and built on that tract an imposing two-story mansion, so impressive for that day that his younger brother, Thomas, was persuaded to make his own house at Prospect Hill an exact copy. The newly created plantation was given the name of Clifton.[41]

William Alston became so fabulously wealthy and influential that all Waccamaw called him "King Billy," and his cousin, Captain William Allston of Brookgreen, had to be content with the title "Gentleman Billy."[42] William of Clifton served as a captain during the Revolution in General Francis Marion's command and later attained the rank of colonel. Sometime before 1792 he dropped an *l* from his last name to distinguish his branch of the family from the others, the Allstons having a tendency toward extreme duplication of first names—much to the confusion of South Carolina in general and its historians in particular. His descendants followed suit—thus the Alstons and Allstons.[43]

William Alston of Clifton in 1777 married Mary Ashe, daughter of Brigadier General John Ashe of North Carolina. Of this marriage, five children were born: Maria; Joseph, who became Governor of

CLIFTON

South Carolina; John Ashe; William Algernon; and Charlotte. Following the death of Mary Ashe, William married Mary Brewton Motte in 1791.[44]

The year 1791 was momentous in other respects: not only did Clifton get a new mistress; it also entertained the first President of the United States. While making his southern tour in 1791, President George Washington came down Waccamaw Neck and spent a night here. The entry in his diary for April 29 reads:[45]

> We left Doctr Flagg's about six o'clock and arrived at Captn Wm. Alstons on the Waggamau to Breakfast.
>
> Captn Alston is a Gentleman of large fortune, and esteemed one of the neatest Rice planters in the State of So Carolina and a proprietor of the most valuable ground for the culture of this article—His house which is large, new and elegantly furnished stands on a sand hill, high for the Country, with his Rice fields below; the contrast of which, with the lands back of it, and the Sand and piney barrons through which we had passed, is scarcely to be conceived.
>
> At Captn Alston's we were met by General Moultree, Colo Washington & Mr. Rutledge (son of the present Chief Justice of So Carolina) who had come out that far to escort me to town.—We dined and lodged at this Gentleman's and Boats being provided, we then next morning.
>
> Saturday. 30th
>
> Crossed the Waggamau to Georgetown by descending the River three miles . . . .

The Dr. Flagg mentioned in Washington's account was Henry Collins Flagg, second husband of Rachel Moore, whose home was Brookgreen.[46] Flagg had invited the President's party in for dinner.

Only a few years after Washington's visit the mansion at Clifton burned. The family saved most of the furnishings from the flames and established their home at Fairfield, leaving Clifton apparently without a main dwelling until Dr. Emerson built one there for his daughter, Margaret Emerson Vanderbilt. The new Clifton house, a place of mansion proportions supported in front by four white columns, was constructed, as nearly as can be determined, during the early or mid-1920's. Mrs. Vanderbilt, who lived there for only a few winters, was staying there in 1927, but since then the big house has been unoccupied.

In Colonel Alston's will, dated November 29, 1838, Clifton was left to his son, William Algernon Alston of Rose Hill. The property remained in William Algernon's estate until it was sold to Isaac Emerson in 1909.[47]

MILL STONE AT ARCADIA

In 1925 Dr. Emerson acquired Bannockburn, a tract of some 3,058 acres, from Ralph Nesbit of Caledonia plantation, to whom it had been given by his father, Colonel Ralph Nesbit. Bannockburn had formerly been the John Izard Middleton place.[48]

Fairfield was the eighth and final plantation to be annexed to Arcadia. After the Clifton fire Colonel William Alston had made his home there, adding two large wings to the small four-room house already standing. His son Charles, to whom he left the property, built a dining room on the side facing the river, bringing the total number of rooms to eighteen. The fine old wall paper and the quaint interior architecture of this residence, the main dwelling at Fairfield, made it outstanding among low-country homes.[49]

Charles Alston married Emma Pringle, member of another well-known coastal family, and they had four children who lived to maturity and to whom the plantation was willed. They were Joseph Pringle, Charles Pringle, Susan Pringle, and John Julius Pringle Alston. Susan outlived all her brothers and left Fairfield on her death in 1921 to her cousin, Mrs. John E. Alston, the former Elizabeth Deas.[50]

During Mrs. Allston's ownership of Fairfield, Henry Ford, the automobile manufacturer, called on her and offered to purchase the plantation's rice mill. Finally persuading her to sell it, he had it removed to his museum at Dearborn, Michigan, where it was re-erected and put into running condition.[51] There at the northern museum it is now possible for people of all sections to observe the old methods employed in South Carolina's bygone rice industry.

Mrs. Allston lived at Fairfield until 1936, when she sold the estate, containing about 993 acres at that time, to George Vanderbilt, who had inherited Arcadia on the death of his grandfather, Dr. Emerson, in 1931.[52] George Vanderbilt died in San Francisco, California June 24, 1961 and was buried four days later in the lovely garden on the south side of the Arcadia mansion. He willed the entire plantation to his only daughter, Mrs. Robert Balding, who has been familiar with this vast property since her childhood. Mr. and Mrs. Balding now make the place their year-round home.

The large Arcadia establishment for many years has been a showplace on plantation tours.

Rossdhu

(Lower Waverly)

THOUGH POSSESSING A NEW NAME, ROSSDHU HAS AN OLD history—one that goes back to the early days of settlement on Waccamaw, for this place until recent years was part of Waverly plantation, established before 1750.[53]

The story of Rossdhu as a separate estate began in the lifetime of two brothers, Francis Williams and St. Julian M. Lachicotte. Upon the death of their father, Philip Rossignol Lachicotte, in 1896, these brothers assumed control of the Waverly Rice Mills, becoming equal partners in the firm P. R. Lachicotte and Sons, and dividing between them the responsibilities and property. Francis kept the main plantation house, located at that time on the northern part of the land, and managed the rice mills, both of which were in the portion still known today as Waverly. St. Julian established his residence on the southern part of the plantation and managed the lumber mill. This southern acreage is the present Rossdhu property.

St. Julian built his house in a grove of magnificent live oaks on a high bluff overlooking the rice fields. Only two hundred yards south in the Rossdhu tract was the site of the original Waverly plantation house. Nothing is known about the design and size of this first dwelling. With its final destruction before the turn of the century, the history of one more home receded completely into the past, to join the other mysteries of the coastland. The house was in such bad condition when Philip Lachicotte purchased the plantation in 1871 that it was not considered worth repairing. Instead, the family lived in a house farther north, one on a hill above the rice mills. The original main residence was taken down, and since then the house on the northern portion, where the Lachicottes made their home, has been called Waverly.

Despite its state of dilapidation, the Waverly slave chapel south of Rossdhu house remained standing until the property's purchase by Captain William Ancrum in 1938. Because of the chapel's unsafe condition, Captain Ancrum razed it to the ground, erecting in its place a large wooden cross bearing an inscription to mark the location. During and even after the days of slavery, the rector of All Saints,

ROSSDHU (LOWER WAVERLY)

Waccamaw, held periodic services in this chapel for the Waverly Negroes.

Still in use at Rossdhu, a short distance from the chapel, is a small box-shaped building that once served as the school, the seat of learning, up to the college level, for all the children who lived in the vicinity. The instructor lived with the St. Julian Lachicotte family during the school term.

Dissolution of the partnership P. R. Lachicotte and Sons followed the death of Francis Williams Lachicotte in 1920, and in 1927 the heirs to St. Julian's share of Waverly sold it to the Rev. James D. Paxton of Lynchburg, Virginia.[54] St. Julian died in 1933. His wife was the former Emma Springs, of Georgetown, whom he had married in 1880. Their children were Philip Rossignol, Mary Jane, Julia Helen, Albert Springs, and Nathan Sampson. Dr. Paxton long had been a nationally prominent minister of the Presbyterian church. He served for a time as pastor of the famous New York Avenue Presbyterian Church in Washington, often called the "Church of Presidents." Under Paxton ownership the house built by St. Julian Lachicotte was the home for many years of the Berkeley A. Woodruff family, now of Columbia.

In 1938 Dr. Paxton sold the property to Captain William Ancrum, a native of Camden, South Carolina, and a retired officer of the United States Navy.[55] The Ancrums completely remodeled the house and landscaped the grounds on an extensive scale. Hundreds of azaleas and camellias were planted beneath the oaks, and the driveway was lined on both sides with gardenias, so that during blooming season the huge masses of color enhance the naturally lovely setting. Such an imposing estate had to be given a new and imposing name—Rossdhu, for a family ancestral home in Scotland.[56]

In 1948 Captain Ancrum sold Rossdhu to a Georgetown real-estate firm,[57] which then sold the 160 acre property to Thomas W. Ruff of Winnsboro, South Carolina in 1951.[58]

Three years later Mr. and Mrs. Ruff changed the residence into an impressive ante-bellum type structure. Gleaming white columns were erected to support a front portico, and wrought iron railings were added to complete the exterior renovation.

The present owners, Mr. and Mrs. Theron E. Hines of Washington, Georgia purchased Rossdhu in October, 1959. They are continuing enthusiastically the development of the grounds, already of notable beauty in this coastal vicinity.

Waverly

FAMOUS FOR ITS LARGE RICE MILLS, WAVERLY, ONE OF the early plantations on Waccamaw Neck, has been in the possession, as far as can be ascertained, of only two families, the Allstons and the Lachicottes.

With the cessation of rice planting, the huge buildings and power machinery of the widely known Waverly Rice Mills were dismantled, leaving the place a silent ghost of the colorful rice era. Here for many years was operated one of the largest rice mills in South Carolina, when planters of the surrounding region used to send their rice to Waverly to be pounded. Though many had threshing mills on their own places, few were equipped to pound the grain. Waverly-processed rice won a gold medal in 1902 at the Charleston Exposition.

The big mills had their beginning under the Allston family, the founders of the plantation. The earliest data indicate that the land which became Waverly was secured partly through grant and partly through purchase by the first John Allston of Waccamaw, one of the two Allston brothers who came from St. John's, Berkeley, for at his death in 1750, this land was part of his holdings. It seems to have belonged in 1794 to William Alston of Clifton, then a captain, though how he came to possess it is not known.[59]

Benjamin Allston, Jr., a son of William Allston of Brookgreen, came into possession of Waverly—though apparently not through inheritance—sometime after 1794, for he is found to be owner of the property in the early 1800's. Benjamin was living at the plantation in 1807 when he made out his will bequeathing the place to his eldest son, Joseph Waties. After Benjamin's death in 1809, his widow, Charlotte Ann Allston, continued to live here until she moved to Chicora Wood in 1819. By then Joseph was old enough to assume control over Waverly himself. Charlotte Ann, a woman of frankness and sturdy constitution, had struggled for ten years to hold the family property intact for her children. In letters to another of her sons, Robert F. W. Allston, later Governor of South Carolina, she frequently let it be known she was not enjoying her task of running the family plantations.[60]

While a student at South Carolina College, now the University of South Carolina, Joseph was appointed a lieutenant in the army. He served in Florida during the War of 1812 and eventually attained the rank of general. Though he married three times, only his third wife, Mary Allan, had any children to live to full maturity. These were two sons, Joseph Blyth and William Allan, who inherited Waverly upon their father's death in 1834. Infants at the time, the two brothers were placed under guardianship of their uncle, Robert F. W. Allston, whom Joseph had made executor of his estate. It was in this way that Robert came to control Waverly until his nephews came of age in 1857.[61]

Waverly soon felt the touch of Robert Allston's good business hand. The plantation was heavily mortgaged when he began operations in 1834, but in a few years he had relieved this financial pressure, and in 1837 he rebuilt the rice mill.[62]

Through one of the strange twists of fate that occur to alter history, complete records of the Waverly plantation and mills for the period of Robert's management survived the War Between the States, while most of his own—those of Chicora Wood among them—which he valued more, were either lost or destroyed after they were sent into the supposedly more secure interior of the state for safekeeping.[63] The not-so-highly-valued Waverly papers have come down largely intact and in 1945 were included in *The South Carolina Rice Plantation: As Revealed in the Papers of Robert F. W. Allston,* edited by J. H. Easterby. The two plantations figuring most prominently in Governor Allston's papers were his own place, Chicora Wood, and the plantation he managed for his nephews, Waverly. Dr. Easterby writes that all of the surviving Allston records have their value, but that the records of Waverly are "outstanding," being "accompanied by vouchers explaining in detail practically every entry."[64] The book contains three pages of tables showing receipts, expenditures, and investments at Waverly from 1834 to 1857.

The two brothers, Joseph Blyth and William Allan Allston, took official possession of their inherited plantation in 1857.[65] They then divided the property—the rice mills, rice fields, and plantation house and grounds going to Joseph Blyth. This is the section still known as Waverly. William Allan received the pinelands later called Woodville.[66] Joseph, the brother who fell heir to Waverly proper, appears

WAVERLY

(Remodeled house not in previous printings)

to have been outstanding for his literary gifts. His cousin, Elizabeth Allston Pringle, writes that he "was a gifted man, clever lawyer and eloquent pleader. His literary talent was above the ordinary; he has written some poems of great beauty; 'Stack Arms' and the longer poem, 'Sumter', deserve a high place in the war poetry of the South."[67]

In May, 1871, Joseph Allston sold Waverly plantation to Philip Rossignol Lachicotte,[68] who was to develop the rice mills to their peak. Philip had served as a second lieutenant in Ward's Light Artillery on Waccamaw during the War Between the States. Once on a special mission with men ordered to take possession of a Federal boat aground in the mouth of Santee River, he accomplished the objective only to be captured by a party from another Federal gunboat. Later he was returned through an exchange of prisoners.[69]

The family history of this little Frenchman was a dramatic story, for Philip was the grandson*of a refugee from the bloody Haitian Slave Rebellion. His grandfather, Pierre Henri Rossignol–La Chicotte, a ten-year-old boy at the time of the violent outbreak in the 1700's, with his mother and four younger brothers and sisters had barely managed to leave the island of St. Domingue in 1792. His father was dead from wounds received fighting off a Negro seige at the family's isolated plantation on the southern tip of the island. An intriguing version of the rebellion, told to a member of the Lachicotte family a few years ago, contends that a squad of slaves had been especially trained and assigned to slaughter this family, the various branches of which seemingly were as prolific on St. Domingue as the Allstons were on Waccamaw.

To an extent the plot met with success. The mother and father, however, had the foresight to send ahead with trustworthy slaves their two eldest sons and some pieces of family silver—all concealed in a vegetable cart headed toward the nearest seaport. Upon the death of her husband, Madame Phillippi Henri Rossignol–La Chicotte with the three smaller children were granted safe escort to the same seaport by a rebel less sanguinary than his fellows. There the mother and five children secured passage on a schooner bound for Philadelphia. Disaster struck once again, this time in the form of an English vessel, which overtook the schooner and relieved the passengers of all valuables. The few small articles Madame Rossignol–La Chicotte had been able to hide from her island captors went to the British. Thus,

**Philip Rossignol Lachicotte was the* son, *not grandson, of Pierre Henri Rossignol-La Chicotte, who arrived in America in 1792 as a refugee from the Haitian Slave Rebellion.*

the family landed in America in 1792 with little more than their lives and the clothes they wore.

The Rossignol–La Chicottes moved to Charleston in 1805. Two years later, the mother left her children in the care of the De Caradeuc family (with whom they later intermarried) and sailed for Cuba—it is believed to enlist aid in going to St. Domingue to recover buried family belongings. She was never heard of again; it is thought that she either was killed on still-turbulent St. Domingue or perished in a sea storm.[70]

Since the family's departure—or at least the departure of certain brothers—from France until the present time, their name has undergone several changes. Originally Rossignol, as the family was known for centuries, the name has evolved into the present form of Lachicotte. As different members became established on St. Domingue, or Haiti as it is known today, each took for the purpose of identification the name of his estate. One branch of the family added La Chicotte, name of their home on the Artibonite River. The family became successively Rossignol de la Chicotte, Rossignol–La Chicotte and eventually LaChicotte and Lachicotte.[71]

Soon after the Lachicotte purchase of Waverly and the formation of P. R. Lachicotte and Sons, modernization and enlargement of the steam-operated rice mills brought them to double their former capacities. Besides improving the threshing mill, pounding mill, and barrel factory already in use, the new owners added a shipyard for the construction of rice barges, a marine railway, a lumber mill, and a post office. To save labor in carrying rice to and from the threshing mill located some distance from the water, the company built a new threshing mill close by the pounding mill. Through the years the mills were improved periodically, and the company even found time to engage in planting rice.

Activity at Waverly reached its zenith during the period from 1871 to 1911—partly because of an increase in the number of industries on the place. Some three to four hundred workers were employed in various phases of the plantation's operation. During the peak of the rice season in those years, Waverly's mills processed an average of two to three thousand bushels of rice per day, placing these mills on a par with the largest in South Carolina. Planters from all over Georgetown County, parts of Charleston County, and part of North Carolina, sent their rice to Waverly to be milled. P. R. Lachicotte and Sons invented and patented in 1875 a "new and improved method" for cleaning rice—the addition of silica powder which enabled polishing machines to clean and whiten the grains in less time and with less breakage than before.[72]

The big mills ceased operating in 1911, as rice planting died in the Carolinas. Rice never since has been planted at Waverly. Dismantling and selling of the machinery, piece by piece, began in 1925. The rotting buildings were also razed, to prevent smoking sight-seers from burning up the entire establishment or injuring themselves in the unsafe structures.

As more and more up-state residents began spending their summers on Pawleys Island, boats came to Waverly, or to Hagley plantation farther south, bringing from Georgetown passengers bound for the beach. From Waverly passengers were transported by horse and buggy to the island. Such travel went out with the days of the mailboat that plied up and down the Waccamaw River. The post office was removed from Waverly in 1923, and in the 1930's the postmark of Waverly Mills was discarded for the name Pawleys Island.

At Philip Rossignol Lachicotte's death in 1896, two sons, Francis Williams and St. Julian, took over as equal partners in the management of the company and Waverly. Francis retained the main residence and managed the rice mills—the part still called Waverly. St. Julian managed the lumber mill and built his own home on the southern portion of Waverly. There were eight other brothers and sisters, but they established homes and businesses elsewhere. These were Isabel Louise Morel, Philip Howard, Louis Claude, Henry Arthur, Clarence Percy, Mary Helen, Walter Betts, and William Fitch. Their mother was the former Mary Jane Miscally of Charleston, who married Philip Lachicotte in 1848.

Francis Williams Lachicotte in 1877 married Lillian Stokes, daughter of the Rev. G. W. Stokes of Vaux Hall, Waccamaw. Francis died in 1920, and his death dissolved the partnership. From him Waverly passed to his children, Marie Hermine, Francis Williams II, Isabelle Morel, Alberta Stokes, and Arthur Herbert. The other three released their interests to Misses Isabelle and Alberta Lachicotte, who still own the plantation and make it their home along with the Arthur H. Lachicotte family.

Waverly became divided for the second time in its history when the heirs to St. Julian's portion sold what is now the Rossdhu estate in 1927 to Dr. James D. Paxton, prominent Presbyterian minister of Lynchburg, Virginia. St. Julian died in 1933. Under a later owner, Captain William Ancrum, the property was named Rossdhu.[73]

Little seems to be known about the original Waverly house except that it stood on a high knoll south of the present Rossdhu residence. The dwelling was in such bad condition when P. R. Lachicotte purchased the plantation in 1871 that it could not be made livable. The home evidently had not been used for many years previously. The Lachicottes moved into a small house on the northern part of the land and enlarged it to the present proportions. This house was made the main plantation residence, and the old house on the hill was dismantled.

All Saints Church, Waccamaw

ONE OF THE EARLY PARISHES IN AMERICA, HISTORIC ALL Saints Protestant Episcopal Church, Waccamaw, is among the few surviving churches of the ante bellum plantation country and seems today to be facing opportunities greater than in times past. While other old churches of the Carolina low country were abandoned with the decline of rice planting, All Saints, Waccamaw, held on with grim determination, remembering an even more disastrous time—its total collapse during the War Between the States. Sacrifice and perseverance mark the history of this colonial parish.

As early as 1739 services were held at a chapel on the site of the present church. John Fordyce, who became rector of Prince George, Winyah, in Georgetown in 1736, was the founder of religious work on Waccamaw Neck, and soon after his coming, the chapel was erected on land given by Thomas George Pawley. On February 1, 1739, John Fordyce wrote, " 'I generally preach at the Chapel of Wackamaw once per annum.' " In this statement was the first mention of a church. For at least thirty years, the little Waccamaw chapel was dependent on the rectors of Prince George for services—an indication that services here must have been few and far between, since Prince George was often without a rector for years at a time.[74]

The Waccamaw settlers were greatly inconvenienced by the necessity of having to cross the river to attend services at Georgetown, and on May 23, 1767, the General Assembly of South Carolina granted their wish for a separate parish. The new parish, All Saints, Waccamaw, carved out of land formerly a part of Prince George, extended from the southern tip of Waccamaw Neck on Winyah Bay to the North Carolina line.[75]

The General Assembly appointed William Allston, Joseph Allston, Charles Lewis, William Pawley, Josias Allston, William Allston, Jr., and John Clarke as commissioners for the building of a new church, chapel, and rectory, and impowered them to buy a glebe and to accept subscriptions. Officers of the new church in addition were expected to care for the poor, promote peace and order, attend to the maintenance of roads and ferries, and supervise elections.[76]

The colonial assembly being none too responsive to the will of the English king, the British government decided to restrict that body by refusing to allow the establishment of new parishes which meant added representation in the assembly. Consequently, the act making All Saints a parish was nullified in 1770. During the Revolution, however, the assembly took action again, and without the veto of a royal governor to hinder them on March 16, 1778, re-established the parish. At the constitutional convention of 1788, Daniel Morall—who had served as a captain under General Peter Horry and General Francis Marion—and Thomas Allston were the delegates from All Saints Parish.[77]

The first rector of All Saints seems to have been James Stuart, who held at the same time the rectorate of Prince George, Winyah, and lived in Georgetown. Stuart took on his duties in 1772 but soon ran into difficulties because of his political views. An equally staunch and outspoken Loyalist, Stuart aroused the ire of his fellow townsmen; they so mobbed and abused him that he resigned in 1777, and for a period no services at all were held at All Saints. The next rector was John O'Donnell, about whom practically nothing is known. His stay was quite short, beginning in July, 1802, and ending the same month two years later. A period of seven years elapsed after his departure before All Saints obtained another full-time minister.[78]

In 1812, when Hugh Fraser accepted the call as rector, church life on Waccamaw began to achieve some regularity. Born in Scotland in 1763, Fraser came to South Carolina in 1793. He first served as rector of Prince Frederick's Parish, until his resignation in 1810. He remained the rector of All Saints until 1817, and in later years whenever the church was without a minister still officiated at services. Mr. Fraser owned and lived at Calais, one of the southernmost plantations formed out of Hobcaw Barony. A prominent figure in church circles, he was a delegate from South Carolina to the General Convention of the Episcopal church in 1814, and in 1815 preached the annual sermon at the diocesan convention. He so impressed his listeners that the request was made that his sermon be published.

The rectorate of Hugh Fraser, says Henry D. Bull, a later rector, in his history of the church, "coincided in time with the episcopate of the Right Reverend Theodore Dehon, the second bishop of South

Carolina. Those were the years of re-awakening in the Diocese, the years of recovery and re-establishment; certainly, this period marks the beginning of a new order at All Saints as in so many other churches in South Carolina."

The cornerstone of the first All Saints church has been badly battered and several dates are missing, but an old copy gives the words of the inscription: " 'The first edifice built of wood was built about A.D. 1798, by Capt. John Allston, was repaired in 1817.' " In a speech of 1817 concerning the erection of a new church in Charleston, Bishop Dehon also commented: " 'In the parish of All Saints, Waccamaw, there has been finished, with a laudable zeal, a very neat and commodious building, which in November [1816] last, was duly consecrated by the name of the Parish Church of All Saints . . . .' " An 1825 map of Georgetown District indicates the existence of the rectory adjacent to the church. A Bible and prayer book presented by Mrs. M. A. Allston at that time are still in the possession of the church.[79]

Because of great distances and difficult roads, the congregation of All Saints found attending services difficult even in the best of weather and next to impossible in bad. A solution was found in the erection of a building at Oak Hill in 1820 to serve residents of the southern part of the parish. The only name ever given this chapel was "The Lower Church." Similarly, in 1854, the All Saints vestry decided to build a church at Wachesaw, some eight miles north of the main church. The cornerstone was laid by the bishop on April 11, 1855, and on May 24, 1857, the building was opened for service under the name of Saint John the Evangelist. This building was called often the "Upper Church." Both branch churches were later destroyed—the lower by fire, the upper by decay. A summer rectory on Pawleys Island for the rector of All Saints was built in 1854 on a 300-foot lot given in 1845 by Colonel Peter W. Fraser. This house still stands and continues to be used for its intended purpose.[80]

Henry Gibbes, a native South Carolinian whose mother was an Allston, was rector from 1819 to 1829. Ordained in 1818 at the age of twenty-one, he came to All Saints for his first charge.[81] Shortly after his resignation, the Waccamaw church secured as its rector a young Englishman, Alexander Glennie, who was to serve as rector for thirty-three years. Francis Marion Weston, wealthy owner of Laurel Hill plantation, in the early 1830's brought young Glennie over from England to tutor his son, Plowden C. J. Weston. Glennie became a lay reader at All Saints and was soon a candidate for the priesthood. He was made a deacon in 1832 and a priest in 1833, and shortly afterwards became rector of All Saints. He served until 1866, when he accepted a call to Prince George, Winyah. There he remained until his death at the age of seventy-six in 1880. He lies buried in the cemetery at All Saints.

Under Mr. Glennie, All Saints flourished. The youthful Englishman received enthusiastic praise for his outstanding work among both planters and slaves in instruction in the Bible and church doctrine. With the willing cooperation of plantation owners, he erected thirteen plantation chapels especially for his ministry among the Negroes. His never-ceasing duties took him to the parish church and neighboring plantations on Sundays and to the more distant plantations during the week. As work at All Saints grew, assistants were employed. Among them was LeGrand Guerry, whose eldest son, William Alexander Guerry, was to become eighth Bishop of South Carolina. In 1851 by instructions from the bishop, Anthony Toomer Porter, a young theological student of Georgetown, reported to Mr. Glennie for work and study. Later A. Toomer Porter became founder of the Porter Military Academy in Charleston.[82]

The cornerstone for the third All Saints parish church was laid December 27, 1843. The first two churches were of wood; the third was constructed of brick and stucco. Funds for the new building came chiefly from Mrs. Benjamin Huger II of Prospect Hill, who had left a sum of $5,000 to the parish, making possible erection of the new church on the same site occupied by the old. The 1843 church, constructed along simple classic lines, was supported in front by four beautifully proportioned columns. Inside the old church, the dark-stained box pews and slave galleries above formed a striking contrast to white plastered walls. Hanging behind the altar were two large bronze tablets, one inscribed with the Apostles' Creed and the other with the Lord's Prayer. Along each side wall were handsome, marble memorial tablets. All Saints was regarded as one of the most beautiful churches in the South.

Members donated many of the fine furnishings installed in the third church. Colonel Joshua J. Ward of Brookgreen presented a pipe organ; Mrs. Francis M. Weston gave a Bible, a prayer book, a chancel chair, a marble font, and a carpet for the chancel desk and pulpit; and

Plowden C. J. Weston contributed the furniture. The church was consecrated by Bishop Christopher E. Gadsden in 1845.[83]

In 1843 a school, All Saints Academy, was started for the instruction of children of the local planters. Located a little south of the church, the school offered English, French, Latin, Greek, music, dancing, and drawing. The schoolmaster worked under the supervision of Dr. Glennie. David Delamer Rosa of New York taught at the school for many years;[84] his descendants have settled in and around Georgetown.

The flourishing parish of All Saints collapsed during the War Between the States. Many families went inland to avoid Federal raids, and all were ruined in the chaos of war. There is no record of any baptisms at the parish church from April 16, 1862, to May 29, 1866, though some were held at Waverly and Brookgreen. Dr. Glennie finally left All Saints in 1862 to act as rector of Prince Frederick's until 1865.[85]

During these years, and those immediately following the war, all Waccamaw was in turmoil as former slaves wrecked plantation and church property. The white people, so vastly outnumbered by Negroes, found life an almost insurmountable difficulty. A report made for the bishop in 1867 on destruction of church property throughout the Diocese of South Carolina sums up in thorough manner the condition of Waccamaw Neck at the end of the War Between the States:[86]

> . . . the prostration of the once flourishing Churches on Waccamaw is complete. They contained more wealth than any other rural parish in South Carolina, or perhaps in the South. There were the homes of the largest rice planters on this continent. Their provision for the temporal and spiritual welfare of their slaves was a standard to other planters. Numerous chapels, built by the proprietors for the use of their people, adorned the estates, where the services of our Church were as well performed as in any other congregation in the land. The faithful labors of their revered pastor, the Rev. A. Glennie, for thirty years had wrought blessed results, aided by the systematic teachings of the planters and their families.
>
> Hundreds of the colored race were communicants of our Church—thousands of colored children recited the catechism, and answered as intelligently as any of their age in Europe or America; and the Lord's work seemed prospering and sure. Alas, for the change in five or six years! Poverty has overtaken these desolated homes; the rice fields, rich as any land between the Mississippi and the Nile, lie desolate; their former laborers can scarcely be induced to work.
>
> Their religious deterioration is painful. They have forsaken the way which they had learned, and taken to themselves teachers of their own

ALL SAINTS CHURCH, WACCAMAW

(Not in previous printings)

> color. Fanaticism and extravagance rule in their religious assemblies to such an extent as to require the aid of the military to keep order and repress violence. There are indications of a return to African barbarism. There are no religious services re-established in the three churches on Waccamaw. The planters are bankrupt; their houses despoiled; their costly libraries torn to pieces or shipped to Northern ports. What good the Lord may work out of the present confusion we know not; but to human eyes the present state of things is only evil to both races which inhabit these beautiful shores.

The slow, laborious recovery of All Saints church in a poverty-strickened era began in 1876 when William H. Barnwell, native of Beaufort, South Carolina, just out of seminary, accepted the call at a salary of $700 per year. He remained until 1885, leaving then for Paris, Kentucky. Custis P. Jones of Baltimore succeeded him the same year and resigned in 1887. He, in turn, was followed by G. T. Wilmer of Virginia, who served the church until 1891. LeGrand F. Guerry, who had served long before as a deacon under Alexander Glennie, came that year to accept the rectorate of All Saints. He also served Prince Frederick's, Peedee. Mr. Guerry resigned in 1904 and was followed in 1905 by Charles E. Cabaniss, a native of Huntsville, Alabama, who died two years later in Georgetown. He is buried at All Saints.[87]

In 1908 John H. Galbraith, a native of Ireland, came from Texas to take the rectorate. He served until his retirement in 1922. Many years of faithful service were given by him to All Saints.

It was Mr. Galbraith's misfortune to be rector when the beautiful third All Saints parish church burned on Sunday night, December 12, 1915. Service had been held that morning and the building long since closed, but a smoldering spark from a defective stove pipe that night burst into flames. A passing Negro spied the fire and dashed for Mr. Galbraith, who managed to save the bishop's chair, the altar vases and cross—memorials to Colonel Ralph Nesbit. The brick building was burned to a shell.

Led by St. Julian M. Lachicotte, senior warden, and Ralph Nesbit II, members of All Saints and interested persons raised funds to rebuild the church. The cause met with such enthusiastic response from people all over South Carolina that construction of a new building, the present All Saints church, was immediately begun under the devoted supervision of Mr. Lachicotte. Consecration by Bishop Guerry of the fourth parish church took place on December 9, 1917, only two years after destruction of the old church—a remarkable feat for this small congregation.[88]

Architecture of the new church followed the same handsome lines of the old—only on a smaller scale. The interior design was quite different, however, slave galleries having been omitted and straight pews installed in place of the old box-type seats. Nor do marble tablets adorn the walls of the present building. Though smaller and less decorative than its predecessor, the church is regarded as one of the fine rural churches of the state.

Dr. and Mrs. Henry Norris of Philadelphia, Pennsylvania, and Rutherfordton, North Carolina, the owners of Litchfield plantation, gave an Italian font of great beauty to the church in memory of their son Charles. Mrs. Norris, a great lover of beauty, also erected the brick fence around the churchyard and donated the handsome iron gates at the entrance. The many and great contributions of the Norrises were a primary factor in restoring All Saints church to its present state. Both Dr. and Mrs. Norris are buried there—near the church they did so much to revive and to make beautiful.

After Mr. Galbraith left in 1922, C. W. Boyd, a native of Union, South Carolina, became rector of All Saints. Considered a brilliant man, he was noted for his excellent sermons. He resigned in 1925 because of bad health and died in 1926. Henry DeSaussure Bull, a native of Stateburg, South Carolina, and descendant of the colonial governor, William Bull, assumed the rectorate in 1925. In addition, Dr. Bull was rector of Prince George, Winyah, Prince Frederick's, Peedee, and minister-in-charge of St. Peter's mission, Snow Mill.[39] He kept up this strenuous pace through many trying years until his resignation in June, 1953, when he accepted a call to the Episcopal church in Barnwell.

The services and devotion given All Saints by Dr. Bull during his many years as rector are beyond measure. He gave much more than duty demanded. His crowning achievement and most distinguished service, perhaps, was his history of All Saints Parish, published in 1948 by the South Carolina Society of Colonial Dames of America. The University of the South at Sewanee, in recognition of his many services to the Episcopal church on all levels, recently awarded him the honorary degree of doctor of divinity. Dr. Bull's presence on diocesan and national committees was a continuous honor in his lifetime, and his election as a delegate to the triennial General Convention of the Episcopal Church was so regular as to become a habit. His personal prestige happily reflected further glory upon All Saints.

Succeeding Dr. Bull as rector of All Saints and of Prince George was Albert Nelson Daunt, a native of Ireland, who came in November of 1953 from Louisiana. Educated at Dublin University, the new rector was in the United States only four years before coming to Waccamaw. Under him the parish made strides toward a still more prosperous life, strides that included renovation and modernization of the church building in 1955.

Succeeding Mr. Daunt was the Rev. Charles A. Robinson, III, who became rector in April, 1959. A native of the Diocese of Upper South Carolina and a graduate of Virginia Seminary, Mr. Robinson has the happy distinction of being the first full-time rector of All Saints in many decades. For his family's occupancy, the winter rectory adjacent to the churchyard was restored to pleasant livability from a state of disrepair in which it had lain for years, untouched, except by vandals.

The future of this colonial parish, again enjoying the services of a resident rector and located near the summer resorts of Pawleys Island and Litchfield Beach, bids well to bring about greater achievements than ever before as the Church reaches out to more and more people.

Litchfield

ONE OF THE VERY OLD PLANTATIONS ON WACCAMAW River, Litchfield has the original mansion, a live-oak avenue, and a broad, sweeping lawn—all of which fit the popular conception of a southern plantation.

The earliest record of Litchfield found today is a plat showing the division of Peter Simons' estate between his two sons, John and Peter. This old document now hangs in the upstairs hall of Litchfield mansion. Peter Simons, an early settler on Waccamaw Neck, apparently died during the year 1794, as the plat is dated November 10 of that year and was drawn up, it reads, at the request of his executors to provide for the specified division of property between the two sons.

The northern half of the Simons property, called Willbrook even at that early date on the plat, was willed to young Peter. It contained at that time 966 acres reaching to the ocean. John inherited Litchfield, the southern half, a narrow portion also consisting of 966 acres and stretching to the seashore. The Simons plat indicates the live-oak avenue leading up to the residence and shows a sketch of the house, a two-story structure with a chimney at each end. The first Peter Simons thus appears to be the man who built Litchfield house, proved by this map to be one of the oldest homes still standing on Waccamaw.

Daniel Tucker of Georgetown bought John's portion of the Simons property sometime between the date of the plat and the year 1796, when Daniel made out his will speaking of his plantation Litchfield, "lately purchased from John Simons."[90]

The Tucker family had taken root in South Carolina in 1722, when John Tucker came to Charleston from Bermuda. His eldest son, Thomas, became a captain during the American Revolution and also served prominently in political capacities. Daniel, the purchaser of Litchfield, was the second son of Thomas. Following in the steps of his father, he also was active in local politics.[91]

On Daniel Tucker's death in 1797, Litchfield, through provisions of his will, went to three of his six sons—John, Daniel Stiles, and

George Heriot Tucker. The other three, Edmund Hyrne, William Hyrne, and Thomas Buchanan Tucker, were left Retreat plantation "on Bullinis Island, formerly the property of Alexander Keith."[92] The eldest of the sons, John Hyrne Tucker (born in 1780) in later years became sole owner of Litchfield—either through agreement with, or upon deaths of the other two brothers.

John Hyrne Tucker was graduated in the class of 1800 from Brown University. He made his home at Litchfield, planting rice there until his death. A muchly-married gentleman, John became the husband of four wives and the father of nine children. He married first, Fannie Caroline Brown of Georgetown, who had no children; second, Elizabeth Blyth Allston, sister of R. F. W. Allston, and had by her two children; third, Susan Harriet Ramsay, daughter of Dr. David Ramsay, the historian, having by her, seven children; fourth, Martha Lavinia Ramsay, sister of his third wife. Martha bore no children.[93]

Quite a colorful personality, John Tucker is the subject of two quaint stories. One, told by J. Motte Alston, recalls a dinner given by the Pee Dee club in honor of President Martin Van Buren during the latter's trip south after expiration of his term of office. Tucker, president of the club, presided. Taking great delight in the vintage of one choice old bottle of wine, the Waccamaw planter called the ex-President's attention to it. He then with relish set about uncorking the bottle. He unfortunately put the corkscrew in too far, and as he was giving one last jerk, up came the cork, screw and all, the tip of the screw striking him on the end of his nose. Blood went trickling down his face. A quick-thinking member tore from his own beaver hat a patch of black fur and stuck it on the wound, and completely unruffled by the incident, Tucker continued his recitation of the wine's merits to Van Buren.[94]

Frederick A. Porcher in his memoirs says that Tucker lived in and for rice, that it was his first and last thought. He also mentions Tucker's pride in his large stock of fine Madeira wine. He tells, too, that John Tucker had an almost superstitious religious faith, firmly believing the Episcopal church to be the only true and safe road to heaven.

When Tucker was seventy-two years old, Porcher recounts, he was very much concerned at the illness of the librarian at the Charleston library where he often went. In Charleston at the time, Tucker

LITCHFIELD

(Remodeled house not shown in previous printings)

stopped in to ask about the librarian each day. One day Porcher told him the man was reported worse.

" 'Ah!' " said Tucker, " 'I suspect our old friend has received his summons for heaven.' "

" 'I think it likely,' " Porcher replied.

Tucker commented, " 'Well, I am four years younger than he. I should be sorry to receive my summons in four years.' "

" 'You are not anxious to go to heaven then?' " asked Porcher facetiously.

" 'Not yet.' "

Porcher smiled and walked a few steps away. Tucker followed, saying, " 'I perceived, sir, that you smiled when I spoke just now, and as I do not wish to be misunderstood on this important subject, I wish to explain myself. I am very anxious to go to heaven, sir, I expect to go to heaven, sir, all my efforts for the last thirty years have been made with a view to go to heaven, but sir, there are many things that attach me to this world. I have a large and promising family of grandchildren, sir, and I am desirous of seeing them settled in life before I am taken away.' "[95]

This delightfully interesting old planter was granted his wish. He received seven more years before his death at seventy-nine in 1859.

Henry Massingberd Tucker, John's son by his third wife, was the next descendant to possess Litchfield. Born in 1831, Henry grew up to become a physician, a profession that did not keep him from acquiring as the same time reputation as a fine hunter and sportsman. He even had pet names for each of his guns. He so consistently proved top marksman in the Georgetown Rifle Club that he soon refused to compete any longer for club prizes.[96]

Dr. Tucker, last of his line to own Litchfield, died in 1904. He had sold the plantation in 1897 to Breslauer, Lachicotte and Company. The company sold its interest in 1901 to Louis Claude Lachicotte, part owner of the firm. Mr. Lachicotte, who at various times during the last years of Carolina rice planting possessed a number of holdings over the county, started at Litchfield in 1890 the first canning factory in South Carolina. Here he packed vegetables and seafood. He sold Litchfield in 1904 to Colonel Ralph Nesbit, who lived at Caledonia plantation just a short distance to the south. Colonel Nesbit sold the property in 1911 to Joshua John Ward, grandson of Colonel J. J. Ward of Brookgreen, and to Arthur Herbert Lachicotte of Waverly.

The Ward–Lachicotte connection sold the plantation in 1926 to Dr. Henry Norris, a noted physician, who restored the plantation house and who also became a generous benefactor of the community. Dr. Norris, a native of Philadelphia, came to Waccamaw from Rutherfordton, North Carolina, where he had built, owned, and operated a hospital, which he later presented to Rutherford County. The Norris estate heirs sold Litchfield in 1942 to Harry Edmond Parker, J. Philip Booth, and Thornwell Hay Parker, prominent businessmen of Sumter, South Carolina.[97]

On assuming possession, Dr. Norris repaired and enlarged Litchfield house, keeping old architecture the same but adding a wing on either side. The residence includes an upstairs ballroom converted into a bedroom and a living room that extends the entire width of the house. One of the most outstanding features is a lovely circular staircase.

Outside, the Norrises put considerable effort into landscaping and developing the grounds. They erected the imposing gateway of wrought iron and English brick at the entrance. The later Sumter owners continued and expanded this program, even landscaping the old street north of the house—once the site of Negro cabins. Banks of azaleas line the avenue and reflection pool on the front lawn.

The only plantation mansion now on Waccamaw to face a live-oak avenue, Litchfield with its stately lines provides an impressive view from the entrance. Both the house and grounds have been handsomely maintained through the years.

In August, 1957 the plantation was sold by the Parkers and J. P. Booth to the present owners, South Carolinians, James B. Moore of Georgetown, E. Craig Wall of Conway, William N. Miller, Jr. of Greenville and Howard Hinman, Jr. of Pine Bluff, Arkansas.

Probably the most significant development in recent years concerning Litchfield has been the growth of its seashore acreage as a beach resort, an investment of the four plantation owners and others. Begun in January, 1959, Litchfield Beach, under the imaginative and vigorous guidance of its developer, Wallace F. Pate of Greenville, already has become an outstanding resort of the Carolina coast.

Here at Litchfield are two interesting and contrasting scenes: one, the old mansion overlooking former rice fields, haunting reminder of the historic rice age; the other, a bustling, modern beach resort community; both, symbolic of the changing times.

Willbrook

THROUGHOUT MUCH OF ITS HISTORY, WILLBROOK HAS been a companion plantation to Litchfield. For a long time the two plantations shared the same owners, but only recently has Willbrook matched Litchfield in importance. It was not until after the War Between the States that Willbrook came into its own, and it is only since the 1900's that it has blossomed into the sizable holding of the present. From its 636 acres of 100 years ago, Willbrook has grown to include the plantations of Oatland and Turkey Hill to the north and covers some 2,400 acres—more than Litchfield.

The earliest record of Willbrook is found on the 1794 plat hanging in Litchfield mansion. Drawn up for the purpose of dividing Peter Simons' estate between his two sons, John and Peter, the plat shows a division of the Simons property into northern and southern sections, the first bearing the name Willbrook, and the second the name of Litchfield. Each contained 966 acres and stretched across Waccamaw Neck to the seashore.

John Simons inherited Litchfield, and Peter Simons got Willbrook. The plat indicates a driveway leading in from the river road to a house on the site between two live oaks where the Willbrook house now stands. The home shown on this early plat evidently was the one that burned during the ownership of Clarence P. Lachicotte, who built the present residence.

Like Litchfield, Willbrook came into the possession of the Tucker family, but at a later date. Daniel Tucker in his will dated 1796 refers to Litchfield, which he had recently acquired from John Simons, but he makes no mention of Willbrook. His son, John Hyrne Tucker (born 1780) probably bought Willbrook sometime after 1800, the date of his graduation from college. At any rate, upon his death in 1959 Tucker left the plantation to his son and namesake, John Hyrne Tucker II.[98]

Young John, born in 1827, came to an early death in 1865. Apparently he died at Plantersville on the Peedee, while saving the life of a Union officer, who was a guest at that time. The sketchy information handed down does not give the nature of the difficulty, the officer was facing. During the War Between the States, Tucker had served as a captain of Tucker's squadron, a unit of the Mounted Men of Winyah, Seventh South Carolina Cavalry, Hampton's Legion. He was graduated in the class of 1848 at Brown University, Providence, Rhode Island, and received his medical doctor's degree from the Medical College of the City of New York.[99]

The next record of Willbrook shows it being transferred in 1877 from Jacob Baum to Elkan Baum.[100] Nothing seems to be known about either of these men, except that Elkan Baum had connections in New York City. After his death, the executors of his estate in 1887 sold the plantation, containing then 766 acres, to Breslauer, Lachicotte and Company.[101] This company for a while had been owned jointly by Louis Claude Lachicotte, Louis Breslauer, and Colonel Ralph Nesbit; but Colonel Nesbit seems to have disposed of his interest in the firm by the time of the Willbrook purchase.

In 1889 Louis Breslauer sold his interest in the property to L. C. Lachicotte, who immediately deeded half interest in the property to his wife, Ella S. Lachicotte, and sold the other half to his brother, Clarence P. Lachicotte.[102] Clarence occupied the old, well-constructed, two-story house on the plantation, and made Willbrook his home, but shortly after his marriage in June, 1895, when he and his bride were preparing to move in, the dwelling burned. He immediately went about building another residence and erected here the house that stands now.

Sometime after 1905 Clarence bought Ella Lachicotte's share, and in 1925 he sold Willbrook, then a place of 636 acres, to the Oatland Beach Company, composed of businessmen from Columbia, South Carolina.[103] Two other plantations, comprising 2,126 acres, Oatland and Turkey Hill to the north, had already been purchased by the company from the Pyatt family. Turkey Hill is believed to have been one of the original Allston settlements on Waccamaw.

The Oatland Beach Company in 1926 sold all three plantations to a wealthy businessman, William S. Ellis of Bryn Mawr, Pennsylvania. Ellis in 1931 conveyed the three tracts to Fidelity Philadelphia Trust Company, which in turn transferred them in 1936 to Jesse Metcalf, a New York City millionaire.[104] Metcalf at the time he purchased these places on Waccamaw was busily acquiring lands also on the Peedee River; before long he became one of the largest property holders in the entire county.

WILLBROOK

(House no longer exists)

After Metcalf's death, his estate in 1941 sold Willbrook, Oatland, and Turkey Hill to the Reed I. West family of Marion, South Carolina. All land in the transaction was included under the name, Willbrook. In 1945 Luther P. Byars of Marion bought the property. Willbrook's most recent sale took place following the death of Luther Byars when it was sold in 1951 to the present owner, J. T. Hunter, of Marion.[105]

Brookgreen

ONCE AN ENORMOUSLY PROSPEROUS RICE PLANTATION, Brookgreen with its lovely old gardens and live-oak avenue has lived on to become the setting of a magnificent array of sculpture and native flora—a gift to the people of South Carolina from Mr. and Mrs. Archer M. Huntington.

Like many of the plantations on Waccamaw, Brookgreen was born under Allston ownership. William Allston, son of the first John Allston of Waccamaw, developed the land into a plantation[106] and constructed here the main house, used not only by his family but by succeeding owners. William (born in 1738) was known as "Gentleman Billy" to distinguish him from his cousin, William of Clifton, who was called "King Billy." A lover of fine horses, "Gentleman Billy" owned one named Trial, that came in first at the Newmarket track in Charleston in 1769, winning a handsome silver bowl still in the Allston family's possession. Later, during the less festive times of the Revolution, William served as a captain in General Francis Marion's command and thus attained his official title. He died in 1781 shortly after becoming sick in camp. Racked with fever, he rode the whole way home, only to collapse upon his arrival.[107]

Perhaps William's greatest distinction is that he was the father of the artist, Washington Allston, whose birth he recorded thus in the family Bible: " 'My son Washington was born on Fryday Night—half after Eleven O'clock the 5th of November, 1779.' " The place was Brookgreen, and the infant's mother was Rachel Moore (afterwards, Mrs. Henry C. Flagg).[108]

Washington Allston's stature as an artist has fluctuated from time to time, but contemporary critics give this son of Brookgreen a position of high importance. Edgar P. Richardson, Director of the Detroit Institute of Arts, writes that Allston "was our first full-scale romantic artist. This in a sentence is his importance for, in so being, he changed the character and enlarged the horizon of American art." Allston was the first American artist to explore the whole range of painting. He did murals, narratives, portraits, landscapes, architecture, animals, and still life. His talents extended to the field of sculpture, which he used in working out studies. He also wrote verse, prose tales, and a novel—all three of which, Richardson says, were "imperfect in form but influential in their time." To these gifts Washington added a reputation as a brilliant conversationalist.[109]

Richardson comments further that Washington Allston's influence on American life was even greater outside his profession. There were "few figures in New England's flowering who were not touched by him in some way," by the quality of his achievements and his life. He gave the young writers of Boston their first awareness of the art of painting. After leaving England for the last time, Allston was elected an associate of the Royal Academy, an honor seldom bestowed upon an American. He was buried in Cambridge, Massachusetts, where he had attended Harvard and had returned to live in his last years. He left a widow, Martha Dana Allston, sister of the New England author, Richard Henry Dana.[110]

Brookgreen was left by Captain William Allston to his son Benjamin, Washington's half-brother and the father of Governor Robert F. W. Allston. Washington inherited Springfield, adjoining Brookgreen, and used it to help finance his art studies abroad. Young Benjamin lost his plantation, however, by going security for a friend—one of many such gestures which proved disastrous to property owners in that day. Apparently two brothers, Robert and Francis Withers, befriended Benjamin in the crisis when the note came due by purchasing part of his holding in 1800. In their honor, Benjamin named his second son, born in 1801, Robert Francis Withers Allston. Joshua Ward, father of Colonel Joshua John Ward, seems to have been the purchaser of all or part of the Brookgreen acreage—probably in 1800, since the tombstone of Colonel Ward states that he was born at Brookgreen in the autumn of that year.

Springfield, the land bought by the Withers brothers, became reunited with Brookgreen in 1847 when Francis Withers died. Francis had willed his Waccamaw property to his niece, Mary Wilkinson. Mary's descendants say that her two uncles thought they had to do especially well by her in property settlements, since she had so long a nose that they feared she would never get a husband. As it happened she did quite well by marrying C. G. Memminger, later Confederate

BROOKGREEN

secretary of the treasury, and when she came into her inheritance in 1847, she sold Springfield to Colonel Ward.

Indications are that the plantation now called Brookgreen once was called Springfield too. The first record of the name Brookgreen appears in the heading of a letter written on November 16, 1843, by Colonel Ward, who thus seem to have given Brookgreen its present name.[111]

One of the largest slave-holders and rice planters in the state, Colonel Ward at one time owned as many as a thousand slaves spread among his various plantations—at his death, the six places of Brookgreen, Springfield, Prospect Hill, Longwood, Alderley, and Oryzantia. He became famous for the big grain rice which he had developed and which earned him a fortune. The Winyah and All Saints Agricultural Society in 1849 awarded him for his accomplishments with this new rice a silver medal now in the possession of the Joshua John Ward family, who live on Waccamaw Neck. In 1851 Colonel Ward was rewarded for his years of service to fellow citizens with the lieutenant-governorship of South Carolina. A year later he died and left, by the provisions of his will of 1848, both Brookgreen and Springfield to his son Joshua.[112]

Real-estate values plummeted during the War Between the States; and in 1870 Dr. Lewis C. Hasell bought Brookgreen plantation from Colonel Ward's estate for the unbelievably low sum of $10,000. Dr. Hasell married Miss Clemence Willet of New York, bringing her down to Brookgreen to live. In the 1880's Mrs. Hasell's brother, Marinus Willet, came from New York with his family to stay at the plantation. Willet purchased the Springfield rice land and built himself a cottage in the original Brookgreen garden. In 1920 Mrs. Willet and heirs to the estate of Mrs. Hasell sold Brookgreen to Dr. J. A. Mood, a prominent physician of Sumter, South Carolina, and sponsor of the Waccamaw Club, formed by a group of sportsmen who used the land for hunting.

Dr. Mood's ownership brought more than hunters to the plantation; more significantly, it brought his daughter, Julia Peterkin, whose frequent visits to Brookgreen made literary history, for she took the general locality of the estate as the setting for three books: *Green Thursday* (1924), *Black April* (1927), and *Scarlet Sister Mary* (1928), winner of the Pulitzer Prize.

The Brookgreen Club took the place of the Waccamaw Club in 1924, and in 1926 W. S. Griffin of Greenville, South Carolina, became owner of the plantation. Griffin's interests later were taken up by the F. M. Credit Corporation, which sold Brookgreen in January, 1930, to Archer Milton Huntington, nationally prominent patron of art, who acquired also the three adjoining plantations, The Oaks, Springfield, and Laurel Hill. The deed covers about 6,635 acres, but Brookgreen Gardens proper takes up only some 45 to 50 acres, centering on the old boxwood garden.[113]

About two miles south of Brookgreen is The Oaks burying ground, which has been partially restored by Mr. Huntington. Following their marriage in 1801, Joseph and Theodosia Burr Alston made their home at The Oaks, willed to Joseph by his grandfather, Joseph Allston of The Oaks. The plantation as such is no more, and the house in which Joseph and Theodosia Alston lived has long since disappeared, leaving the land once again a wilderness. But the still-intriguing tale of the ill-fated Theodosia lives on in romantic minds.

A gifted and ambitious man, Joseph won the governorship of South Carolina in 1812, but the year which began so brilliantly for him ended in tragedy. On June 30 of that year, his ten-year-old son and only child, Aaron Burr Alston, died of malarial fever while the family were staying at The Castle on Debordieu Beach, and Theodosia was never able to shake off the grief caused by loss of her only child.

At the persistent urging of her unhappy father that she visit him, Theodosia left The Oaks on December 30, 1812, for Georgetown, where she boarded a schooner, "The Patriot," bound for New York. The ship never reached its destination nor was any word ever heard again of the crew or passengers. Serious consideration has not been given either by the family or by historians to the tale originating from Nag's Head, North Carolina, that the schooner was captured by pirates and all aboard forced to walk the plank. "The Patriot," really a pilot boat fitted out as a privateer after war began, was carrying her guns below deck as ballast. The breaking loose of these guns would have been enough to cause the ship to founder during the heavy gales known to have whipped the Carolina coast through the first days of January, 1813.[114]

The tombstone over the graves of Governor Joseph Alston and his small son in The Oaks burying ground concludes the story of three tragic lives:

Erected to the Memory
of
Joseph & Theodosia Burr Alston
and their Son
Aaron Burr Alston
This last died in June 1812 at the age of 10 years
and his remains are enterred here.
The disconsolate Mother perished a few
Months later at sea.
And on the 10th Sept. 1816 died the Father
when little over 37 years of age, whose remains
rest here with the Son's
The loss of this Citizen was no common one to
the state. To its service he devoted himself from
his early years.
On the floors of its legislature he was
distinguished for his extensive information &
his transcendent eloquence in the chair of the
House of Representatives for his impartial
correct decisions & everywhere, he was
distinguished for his zealous attachments to
republican principles.
in the capacity of Chief Magistrate of the
state when both honour & the responsibility
of the trust were heightened by the
difficulties and dangers of the war of 1812
he by his indomitable activity & his salutary
measures earned new titles to the respect & the
Gratitude of his fellow citizens.

This great man was also a good one.
He met death with that fortitude with which
his ancestor did, from whom he received his
name & this estate & which to be found
in the good hoping to rejoin those whose loss
had left in his heart an aching void that
nothing on earth could fill.

Courtesy of Mrs. John E. Allston

THE CASTLE ON DEBORDIEU BEACH, WHERE AARON BURR ALSTON DIED

When Mr. and Mrs. Huntington purchased Brookgreen in 1930, their first thought was to use it as a winter home, but they soon conceived the idea of Brookgreen Gardens. The development of the idea is told in Mr. Huntington's words:[115]

> Brookgreen Gardens is a quiet joining of hands between science and art. The original plan involved a tract of land from the Waccamaw River to the sea in Georgetown County, South Carolina, for the preservation of the flora and fauna of the Southeast. At first the garden was intended to contain the sculpture of Anna Hyatt Huntington. This has gradually found extension in an outline collection representative of the history of American sculpture, from the nineteenth century, which finds its natural setting out of doors. It is not an experiment station nor a research plant. Its object is the presentation of the natural life of a given district as a museum, and as it is a garden, and gardens have from early times been rightly embellished by the art of the sculptor, that principle has found expression in American creative art . . . .

The best brief account of Anna Hyatt Huntington's world importance as a sculptor is found in the pamphlet published by Brookgreen Gardens. It reads in part:[116]

An equestrian statue of Joan of Arc, for which the sculptor was decorated with the purple rosette by the French Government, was erected on Riverside Drive, New York, in 1915, and replicas were placed at Blois, France, Gloucester, Massachusetts, and San Francisco. It was awarded the Rodin Gold Medal at Philadelphia and the Saltus Medal for Merit at the National Academy of Design. She made a wall statue of Joan of Arc for the Cathedral of Saint John the Divine in 1922. In that year her *Diana of the Chase* received the Saltus Medal. She married Archer M. Huntington in 1923. The equestrian statue of the Cid Campeador was erected at Sevilla in 1927; replicas are at San Francisco, San Diego and Buenos Aires. The works which she made for the grounds of The Hispanic Society of America, New York, from 1927 to 1937 are a replica of *The Cid Campeador,* two flagpole bases, two lions, a red stag, a red doe and fawn, and four animal groups. *Bulls Fighting* won the Shaw Prize at the National Academy of Design in 1928 and *Greyhounds Playing* the Widener Gold Medal at the Pennsylvania Academy in 1937. She designed a monument to Collis P. Huntington erected at Newport News. The Spanish decoration of the Grand Cross of Alfonso the Twelfth was given her in 1929. In the next year she was awarded a gold medal for distinction in sculpture by The American Academy of Arts and Letters. In 1932 she was given an honorary degree of Doctor of Fine Arts by Syracuse University, and in 1933 she was made an officer of the Legion of Honour.

More than a patron of the arts, Archer Huntington was also a poet, a noted linguist, and one of the foremost authorities in this country on Spanish culture. He died December 11, 1955.

A third person who has played a large part in the development of Brookgreen is Frank G. Tarbox, a native of Georgetown, South Carolina, the director and curator of flora. He has been at Brookgreen since the inception of the gardens and not only collects and plants the native flora but has the daily responsibility and management of the entire establishment.

Robert A. Baillie, curator of sculpture and a trustee of the gardens, is another person of importance to Brookgreen. Born in Scotland, Baillie comes from a family of stone-carvers and has become one of the world's most noted men in that art. "When it was no longer possible for Mrs. Huntington to carry out all her own work, she turned to Baillie as the best possible interpreter of her models." He is responsible for the location and placement of statues in Brookgreen.[117]

Still another person closely associated with the gardens for many years is Mrs. Genevieve Willcox Chandler, whose colorful personality has charmed visitors at the Brookgreen museum of small statuary, where she has been hostess for almost as long a time as the estate has been opened to the public. She is a well-known authority on low-country folklore and is a *raconteuse* of Gullah tales.

Brookgreen Gardens was incorporated in 1931 and in 1935 was endowed with a million dollars. These magnificent gardens are open to the public free of charge throughout the year.

An additional service to South Carolina was the decision by the Trustees of Brookgreen Gardens to lease in October, 1960 on a long-term basis most of the seashore property to the state for development as a state park. By the autumn of 1961, work on the project was well underway, and this beautiful tract of beach and woodland was taking shape as Huntington Beach State Park. It is a fitting tribute by the state to name the park for the founders of Brookgreen Gardens, who gave so splendidly of themselves and their resources to South Carolina throughout the many years after their purchase of Brookgreen plantation.

Wachesaw

WACHESAW BEARS THE STAMP OF TWO VASTLY DIFFERENT civilizations—that of the American Indian and that of the South Carolina rice planter. The residence at Wachesaw is located directly upon an old burial ground used centuries ago by the Waccamaw Indians, who inhabited the long peninsula later named for them.

The name of the plantation is variously interpreted. Some say that "wachesaw" means "happy hunting ground," while others claim that the correct meaning is "place of the great weeping." It all depends, it would seem, on one's point of view.

It is supposed that the plantation was once owned by an Alston or Allston—which one is not definitely known.[118] It *is* known that by 1825 the property was in the possession of Rev. James L. Belin, and it is possible that he either inherited or purchased Wachesaw from an Alston or Allston relative.[119]

Born in All Saints Parish in 1788, Belin was from one of the first families to settle on Waccamaw Neck. He became a wealthy man and a distinguished clergyman of the Methodist church. His estate is still managed by the South Carolina Methodist Conference, for he left a sum of money for furthering mission work and particularly for the maintenance of the Methodist church at Murrells Inlet, which was his pastorate for many years and which is named Belin Memorial Methodist Church in his honor. He died in 1859, after falling from a cart while driving along the Wachesaw road.[120]

Mr. Belin apparently gave Wachesaw and The Hermitage to his nephew, Dr. Allard Flagg.[121] At any rate, by some means Dr. Flagg did come into possession of both Wachesaw and the accompanying summer home, a lovely place known as The Hermitage, the two comprising 937 acres. These two pieces of property were bought and sold together for many years.

At the height of pre-war church activity (and probably after Dr. Flagg, an Episcopalian, became owner) the vestry of All Saints Church, Waccamaw, decided on December 24, 1854, to build a smaller church at Wachesaw to serve the inhabitants of upper Waccamaw Neck, who found traveling the long, sandy road to the main church difficult. Such a building had been erected earlier for residents on lower Waccamaw. The cornerstone for this church at Wachesaw was laid by the Rt. Rev. T. F. Davis, Bishop of South Carolina, on April 11, 1855, and on May 24, 1857, the building was opened for services. It was consecrated under the name Saint John the Evangelist, although it was often called simply the Upper Church.[122]

Dr. Flagg, along with Francis W. Heriot and Plowden C. J. Weston, served on the building committee for this church, yet it is said that Dr. Flagg after the War Between the States used the lumber from the building to erect a summer cottage on the seashore. The Rev. Alexander Glennie, rector of All Saints, left Waccamaw during the war, so the story goes, and the Church of St. John the Evangelist went unused and soon fell into bad condition. Noticing its state of dilapidation, Dr. Flagg decided to tear up the building and to use the lumber for a cottage. Local Negroes warned him sternly that "e' jes' ain' fittin' fo' mek house out chuch," but the immovable doctor proceeded with his plans. Soon after the walls of the cottage were put up, a storm blew them to the ground. Undaunted, Dr. Flagg gathered the scattered lumber and rebuilt. Another storm blew the house down. This second outburst of the elements not only ruined the old lumber and succeeded in discouraging Dr. Flagg, but also marked the end of his cottage and the Church of St. John the Evangelist. Only the entrance steps of the church are still to be seen at the original site on Wachesaw. The silver chalice and paten used there and engraved with the name St. John the Evangelist are now in the possession of All Saints Church, Waccamaw.[123]

Dr. Flagg married Penelope Bentley Ward, a daughter of Colonel Joshua John Ward of Brookgreen, and they had two children, Alice and Allard Belin Flagg. Mrs. Flagg died early, leaving her husband to manage her children's share of Colonel Ward's estate until they should come of age.[124] During the last years of his life Dr. Flagg and his daughter Alice lived at The Hermitage. The doctor died in 1901.[125]

Before his death, Dr. Flagg on December 18, 1896, conveyed Wachesaw and The Hermitage to Hesse E. Belin, who deeded them to Alice Flagg on June 4, 1897. In 1904 Alice sold both places to Samuel Sidney Fraser of Georgetown. Fraser's estate then sold them in 1905 to Robert Ernest Beaty, member of an old Georgetown family.[126]

WACHESAW

(House no longer exists)

Beaty sold Wachesaw and The Hermitage in 1910 to Clark A. Willcox of a prominent Marion, South Carolina, family. He was the father of Mrs. Genevieve Willcox Chandler, who has been hostess at Brookgreen Gardens for many years. In 1930 Mr. Willcox sold Wachesaw, consisting of 674 acres, to the present owner, William A. Kimbel of New York City, but kept The Hermitage, which is now the home of Clark A. Willcox, Jr.[127]

Discovery of the burial mound used long ago by the Waccamaw Indians was made soon after the Kimbels came to Wachesaw, as workmen were excavating a portion of the river bank to construct a hunting lodge. A skull filled with beads of the type commonly used by European traders was found only a few feet below the ground's surface, and some Venetian beads uncovered are thought to have come from the Spanish, who centuries ago settled for a short time on this part of the coast.[128] Two poems, "Beads" and "Waccamaw Mound," by Dr. Chapman J. Milling, the poet, writer, and authority on Indian history, were inspired by discovery of the trade beads at Wachesaw.[129] Some five or six skeletons were unearthed from land about the cabin in May, 1930, and in February, 1937, several burial urns were uncovered. In circle formation the graves contained skeletons crouching beneath large burial pots. The bodies were arranged in this position, ready to spring to the hereafter, say authorities on Indian lore. Though the Waccamaws were regarded as a war-like tribe, the graves indicated that no massacre or epidemic had occurred, for great care had been taken in burying the dead. From time to time bits of pottery have been found scattered along the bank. Among the relics discovered at the plantation are spoons, bracelets, corn-grinding stones, and a flint sharpener. Some of these form part of the Indian collection at the Charleston Museum.[130]

The original Wachesaw house, which stood in the grove of oaks on the approach to the river, burned about 1890. The present, modern-style house on the river bluff was built by William Kimbel in 1940. Included now in Wachesaw property is the old plantation just to the south known as Richmond Hill. Kimbel acquired the tract in 1930 from J. Fred Grant, the property consisting at that time of 480 or more acres.[131] Richmond Hill in 1851 belonged to Dr. John D. Magill.[132] Still to be found in the wooded areas here are remains of an old Revolutionary fort overlooking the river. Outlines of trenches and lookout points are still visible.

Considerable prestige has been given Wachesaw by the outstanding achievements of the current owner. Mr. Kimbel's most recent assignments for the United States Government have been as chief American delegate to sessions of the Economic Commission for Europe and as adviser representative for the State Department in meetings of the United Nations Economic and Social Council (of which the Economic Commission for Europe is a unit) at Geneva in June, 1954.

Mr. Kimbel was for many years president of A. Kimbel and Son, a firm dealing with design, architectural construction, and the manufacture of woodwork and paneling. During World War II he was assistant to General William J. Donovan, Director of the Office of Strategic Services, and in 1948 and 1949 he served as Director of the Anglo-American Council, an organization formed to help solve England's need for increased production and export. Kimbel came into political limelight in South Carolina during the Presidential campaign and election of 1952 as an early supporter of Dwight D. Eisenhower. He later served for a time as representative and adviser to the administration on policy and patronage in the state.

Mr. and Mrs. Kimbel have made Wachesaw their year-round home. The plantation in recent years has been included in various plantation tours.

Waccamaw Neck Addenda

Hobcaw Barony

The major development at Hobcaw Barony since the days when the renowned presidential advisor Bernard M. Baruch used it as a vacation home has been the establishment here of the Belle W. Baruch Foundation. The foundation's stated purpose is to use this property for "teaching and/or research in forestry, marine biology, and the care and propagation of flora and fauna in South Carolina, in connection with colleges and/or universities in the state of South Carolina." This generous gift to South Carolina by the very farsighted Belle Baruch has been of monumental importance to the state and came before the preservation of flora and fauna and the proper environment for marine life became the popular causes they are today.

The chronology of events resulting ultimately in the Belle W. Baruch Foundation facilities and program as they are in 1988 began, really, back in 1936. During that year Belle Baruch bought from her father Bernard M. Baruch the portion of Hobcaw Barony that had become known through the years as Bellefield plantation. By 1939 Belle had constructed here a two-story residence where she lived for the remaining years of her life. In 1956 Belle purchased the rest of Hobcaw Barony from her father, who by this time was less active than in former years and was staying primarily at a smaller country place near Kingstree, South Carolina.

Upon Belle Baruch's death in 1964, the trust providing for the foundation came into being. She had specified in her will that the organization be named for her father. However, Bernard Baruch outlived his daughter and insisted that the newly-created foundation bear her name. So it does.

The program at Hobcaw Barony has grown in phases. In 1968 the Belle W. Baruch Forest Science Institute of Clemson University was created "to conduct teaching and research in forestry and related sciences." During the following year, the Belle W. Baruch Institute for Marine Biology and Coastal Research of the University of South Carolina "was established to conduct teaching and research in the marine sciences and related fields." Construction was finished in 1973 on the Belle W. Baruch Field Laboratory of the University of South Carolina. In 1978 the Kimbel Living and Learning Center was built near the entrance to Hobcaw Barony and consisted of a conference center and three cottages. Two dormitories were erected at the complex in 1982. Also built in 1982 was the Bellefield Nature Center, which is located at the entrance to the barony. It is open to the public.

The first floor of Bernard Baruch's imposing residence is open to the weekly tours of the property. The second floor houses offices of the Belle W. Baruch Foundation and allied organizations.

Bellefield

Conveyed to Belle W. Baruch in 1936 by her father Bernard M. Baruch, Bellefield rejoined Hobcaw Barony in 1956 when Belle purchased the remaining barony acreage from her father.

From the time Belle completed the present Bellefield house in 1939 until her death in 1964, she lived here. The house is still maintained as a private residence.

For other developments concerning Bellefield, see "Hobcaw Barony."

Arcadia

Like most of Georgetown County's plantations, Arcadia has undergone change in recent years. The first change occurred in April, 1970, when Lucille Vanderbilt Balding, who had inherited the place in 1961 from her father George Vanderbilt, sold 2,750 acres to her future husband Wallace Pate. This acreage comprised the southern part of DeBordieu Beach and extended from the Atlantic Ocean westward to Highway 17. Wallace Pate soon afterwards developed the area into a luxurious resort residential community called DeBordieu Colony, now well-known along the southern coast. The next change was the sale on July 15, 1985, by Wallace Pate of The DeBordieu Corporation, encompassing everything at DeBordieu Colony, to the North Inlet Corporation, owned by developers Joseph A. Edens and H. Dan Avant of Columbia, South Carolina. On the same day, the North Inlet Corporation purchased an additional 74.34 acres, including beachfront footage, from Lucille Vanderbilt Pate.

The most recent transaction was the sale in November, 1985, by Lucille Pate of more land from Arcadia to another Edens and Avant entity, this time the Prince George Joint Venture Corporation. This tract of 1,977.77 acres included land from the Waccamaw River to the ocean and came off the northern part of Arcadia. Some 368 acres were from the river to Highway 17, and 1,604 acres were from Highway 17 east to the ocean.

Despite the foregoing land sales, Arcadia is not in danger of disappearing. The estate in 1988 still consisted of 8,000 acres, a substantial holding by any standard.

Wallace and Lucille Vanderbilt Pate were married on February 18, 1971, and have made their home since then at Arcadia. The handsome eighteenth century mansion here, the original Prospect Hill plantation house, and the lovely gardens, also of eighteenth century vintage, have always combined to make the plantation a showplace. Today, Arcadia still ranks as one of South Carolina's most spectacularly beautiful plantations.

Arcadia was placed in 1978 on the National Register of Historic Places.

Rossdhu
(Lower Waverly)

Mr. and Mrs. Theron Hines of Washington, Georgia, who purchased Rossdhu in 1959 from Thomas W. Ruff, ultimately made significant changes in the place. First of all, they dropped the name *Rossdhu*, which was a new name anyway. Captain William Ancrum, after buying the property in 1938, started calling it Rossdhu for an ancestral home in Scotland. The Hineses began in 1961 to use the name *Lower Waverly* in an effort to return at least partially to the property's origin as part of Waverly plantation. Next, they subdivided and sold off most of the tract. Homeowners who built on the southernmost part of the bluff overlooking the river chose to return to the use of *Rossdhu* as the name of their area.

On January 16, 1974, Mr. and Mrs. Philip H. Prince of Spartanburg, South Carolina, purchased the main house and the surrounding grounds consisting of sixteen acres from Mr. and Mrs. Hines, who then constructed a smaller house just a little to the south. Since 1985

Mr. and Mrs. Prince have made Lower Waverly their permanent home.

The house and grounds here are quite lovely, as they have been for some years, and present one of the most outstanding views on Waccamaw Neck.

Waverly

Alberta Stokes Lachicotte, who survived her sister Isabelle, died in 1977, leaving Waverly to her nephew Arthur Herbert Lachicotte, Jr., and to her niece Alberta Morel Lachicotte Quattlebaum. These two heirs, who had grown up here, then developed the back part of the plantation into a residential subdivision called Waverly Mills. They also divided the remaining portion between them except for parceling out several tracts.

Alberta Quattlebaum came into possession of the 32.8-acre section containing the old Waverly house used by the Lachicotte family since 1871. In 1983 she sold six acres in an outlying field to Philip H. Prince, owner of Lower Waverly, leaving the present number of 26.8 acres around the residence. She and her husband Jesse Clifton Quattlebaum, Jr., then moved the house about one hundred yards northwest to its present site overlooking the water and rice fields. Following the completion of restoration work on the house in August, 1984, Mr. and Mrs. Quattlebaum and their daughter Rebecca made Waverly their year-round residence.

Like most of the plantations on Waccamaw Neck, Waverly has undergone drastic change. Out of the large number of old outbuildings once on the premises, only three now remain. The place serves as an example of the tremendous change wrought by the resort economy.

All Saints Church, Waccamaw

The story of the colonial parish of All Saints, Waccamaw, in recent years has been, for the most part, a very happy chronicle of tremendous physical and spiritual growth. The autumn of 1988 found this Episcopal congregation moving at last into its new sanctuary, the fifth edifice, marking the fulfillment of more than a decade of effort. Without question, this old parish is at the present time better off than ever before and is able to minister to the whole Waccamaw area more effectively than at any previous time in its two-hundred-twenty-one-year-old history.

The rectorate of the Reverend Charles A. Robinson, III, from 1959 to 1963 was this century's first spurt forward at All Saints. It was the first time since 1925 that the parish had a full-time rector. The Rev. Mr. Robinson laid the foundations for new growth as he steered the parish in that direction.

During the tenure of the next rector, the Reverend Allen C. Mustard, from 1963 until 1970, the parish progressed to the point where it began to think seriously of a larger church building to accommodate the large summer attendance. However, it was not until the rectorate of the Reverend John H. Templeton from 1971 until 1977 that any construction actually took place.

The building program took place in three phases. The first was a new parish house completed in 1976. The structure had actually been almost completed in 1975 when lightning struck it, causing it to be gutted by fire. The second phase was an educational building consisting of classrooms, and this was finished in 1983. Much of the planning for the educational building went on during the time the Reverend Randolph Charles, Jr., was rector *locum tenens* and during the rectorate of his successor, the Reverend Frederick Lindstrom, Jr., from 1978 until 1981.

The present rector, the Reverend Charles H. Murphy, III, who came to All Saints Church in August, 1982, has provided the dynamic leadership resulting finally in completion of the new church, an imposing architectural blend of Greek and contemporary styles. The $1,900,000 structure has a seating capacity of 450 to 500 persons. The conceptual design—by the late Philip Ives, a noted church architect of Greenwich, Connecticut—contains provisions for transepts that can be added to increase seating capacity to 600 persons. Architect Dennis Donahue of Charleston drew the working plans for the building. The old church will continue to be used for small services.

The first service ever held in the fifth edifice was a celebration of the Holy Eucharist on September 28, 1988. It was the special opening for a diocesan-wide leadership conference and was attended by some 450 people. The first regular parish service to take place in the building was a celebration of the Holy Eucharist on Sunday, October 2, 1988. Dedication and consecration took place on Sunday, December 18, 1988.

The South Carolina Lowcountry is dotted with handsome old church buildings now standing unused in the countryside. They are victims of the changing economic times that ended the great rice plantations along the coast. All Saints Parish is a wonderful exception.

In 1991 the old church and St. John's Chapel (formerly Cedar Grove Plantation Chapel) were placed on the National Register of Historic Places.

Litchfield

What was once a rice plantation with acreage extending from the Waccamaw River to the Atlantic Ocean is now a complex of resort developments stretching from the river to the sea. This rather dramatic change over the last thirty or more years is the overriding aspect of the recent history of this eighteenth century plantation.

The lovely eighteenth century house is still here and is beautifully maintained. It still stands at the head of the old live oak avenue, also well maintained. The grounds, too, are still nicely landscaped despite the changed atmosphere from one dwelling to many.

The gradual selling away of land began in the 1950's with the beach tract, now known as North Litchfield Beach. Afterwards came the sales of the area now called Litchfield-by-the-Sea and the area just south of that now named Litchfield Beach. The middle woods section of the plantation was sold and became two country club developments—first the Litchfield Country Club and then the River Club. Each has a club house, a golf course and a residential section. A. Foster McKissick of Easley, South Carolina, board chairman and chief executive officer of The Litchfield Company, has been the leading figure in all this development. He is to be commended for the consistent attractiveness of his developments which preserve to a remarkable degree the natural beauty of the land.

The river portion of the plantation west of River Road is all the land still attached to the old house in 1988, and residential development has come even to this portion. There are resort condominium complexes on both sides of the main avenue. The old house is reserved for special use by the present owners and by residents in the river portion of the plantation.

Some of the change at the plantation began during the Parker-

Booth ownership and continued during the ownership of the investing group of James B. Moore, Craig Wall, William N. Miller, Jr., and Howard Hinman, Jr., who bought the place in 1957 from the Parker and Booth families. After selling off some acreage, the Moore, Wall, Miller, Hinman group sold Litchfield in 1963 to Clay Price of Greensboro, North Carolina. Upon Clay Price's death in 1966, the plantation, consisting then of 625 acres, came into the hands of his brother Julian Price, II, and his sister Louise Price, married at the time to Young Smith. In 1969 the property was put under the name of the Litchfield Plantation Company. Louise Price, now married to Donald Parsons and residing in Palm Beach, Florida, is still the president and principal stockholder of the Litchfield Plantation Company.

The Price family in 1970 had the old plantation house renovated and redecorated. In an effort to return the house to its original proportions, the owners ripped away the back dining room wing, an addition constructed by Dr. Henry Norris in the 1920's. The interior was then elaborately redecorated by Otto Zenke, an interior designer of Greensboro.

The present owners then in 1971 began the residential development of the plantation. There are about twelve single family homes and forty-three condominium units on the place at the present time. An outbuilding called a carriage house is now a private restaurant.

Litchfield was one of the first plantations on Waccamaw Neck to undergo the drastic change now so prevalent in the area.

Willbrook

The Willbrook tract—still consisting for the most part of three old plantations, Willbrook, Oatland and Turkey Hill—is the most recent Waccamaw River plantation set up for development.

Comprising some 2,305 acres, the tract was sold in 1984 by the Hunter family of Marion, South Carolina, to the Litchfield-by-the-Sea Partnership, Inc., which plans to construct over a period of years a huge resort development here. When completed, the complex will have three golf courses and 3,800 homes and will rival the population of Georgetown. The first golf course has already been completed. Also completed is the dismantling of the old Willbrook plantation house. For the time being, the site is to be maintained as a green, open area.

J.T. Hunter of Marion had purchased Willbrook in 1951 and had kept it basically intact. After Mr. Hunter's death in 1970, his heirs, Hattie Costa Hunter King, J. Thomas Hunter, Jr., Adelle Hunter West and Dorothy Hunter Thames, continued to hold the place intact until the 1984 sale. The new owner, Litchfield-by-the-Sea Partnership, is a joint venture of the Marathon Oil Company and of The Litchfield Company headed by A. Foster McKissick.

At the present time, the property is undergoing much alteration. A new, landscaped entrance is now on Highway 17. To say the least, the future appears to hold interesting things for Willbrook.

Brookgreen

Since its incorporation in 1931, Brookgreen Gardens has been one of South Carolina's greatest cultural assets and in recent years has become even more so. Even in the Colonial period, Brookgreen was an outstanding plantation and continued to be outstanding until the plantation era ended in the early twentieth century. It is appropriate that the place today ranks as "the leading institution in the world for the collection and exhibition of American sculpture," a claim made by Joseph Veach Noble, president of Brookgreen Gardens.

From the day Brookgreen Gardens was first opened to the public in 1932, emphasis has been on growth and expansion. For many years, Mr. and Mrs. Archer M. Huntington, the founders, planned all development. After Mrs. Huntington's death in 1973, planning became fully the work of the Brookgreen Board of Trustees and the director. Frank G. Tarbox, who had been here from the beginning of the Gardens, retired as director in 1963 and was succeeded by his nephew, Gurdon L. Tarbox, Jr., the present director. Gurdon, like his uncle, had specialized training to equip him for the demanding Brookgreen position. He earned a bachelor of science degree in forestry from Michigan State University and a master's degree in forestry from Purdue University.

Under the guidance of the trustees, Brookgreen Gardens embarked during the mid-1960's on a new phase of long range planning.

The first new milestone was the use of an admission fee, beginning in 1967. The endowment from the Huntingtons had become painfully inadequate for operation of the gardens. An outgrowth of the admission policy was the establishment of the ever growing membership program. The resulting financial improvement has made possible expansion in various spheres of activity.

The present Visitors Pavilion was completed in 1968. Removal of the wildlife park from its central location to a more northerly site was accomplished in 1977. Further development of the park is an ongoing process. Another major change was a new road system completed in 1980. A year later Brookgreen Gardens took time to observe its fiftieth year of existence by publishing the very handsome book entitled *A Century of American Sculpture: Treasures from Brookgreen Gardens.*

Another important milestone in Brookgreen's growth was the completion in 1983 of the Education Center near the new wildlife park. This center made possible the development of a comprehensive educational curriculum serving both children and adults. Brookgreen Gardens is an accredited museum and as such "is primarily an educational institution," states Gurdon Tarbox. The center sponsors on a year-round basis classes, workshops and publications dealing with the art, flora and fauna of the gardens.

The next project was the creation of the Fountain of the Muses Garden south of the original sculpture gardens. Dedicated in 1984, this new garden has as its focal point the impressive Carl Milles work, *Fountain of the Muses,* acquired from The Metropolitan Museum of Art, New York City. Here, too, further development continues.

The most recent addition to the whole complex is the Wall Building, which houses the administrative offices for the operation of Brookgreen Gardens. Completed in 1986, the structure is named in memory of a prominent and generous benefactor, the late E. Craig Wall of Conway, South Carolina.

Simultaneous with construction at Brookgreen in recent years have been the numerous acquisitions of individual works of sculpture by American citizens. Additions to the art collection are placed not only within the garden walls but also in various areas beyond the walls. Paul Manship's work, *Time and the Fates of Man,* a major sculptural accession for the arboretum beyond the Visitors Pavilion, is among the significant number of pieces now spilling into the unwalled grounds.

An exciting, new direction into which Brookgreen has ventured during the last several years is the field of the performing arts. The satisfactory sponsorship of symphonic and choral concerts and of an operatic production at the Gardens has provided another facet of cultural opportunity to the region.

The potential of the beautiful land here is unlimited. Those entrusted with the future of the Gardens plan for Brookgreen to "continue to grow and to serve, not just South Carolina, but the whole eastern region of the United States as a place of inspiration, education and recreation."

Brookgreen Gardens was placed on the National Register of Historic Places in 1978. The listing also includes Laurel Hill plantation and The Oaks plantation, parts of the present Brookgreen tract.

Wachesaw

Wachesaw is yet another Waccamaw River plantation now undergoing development as an exclusive country club residential community.

The deaths of long-time owners Mr. and Mrs. William A. Kimbel triggered this drastic change at Wachesaw. Buying the place in 1930, the Kimbels had made it their year-round home during most of those years. In the 1970's, they took keen interest in broadening local cultural opportunities and made generous financial contributions for that purpose to selected educational institutions of the area. The philanthropic gifts of this very community-minded couple have resulted in enduring bequests for all of Waccamaw Neck. William Kimbel died in December, 1980, and his wife L. Maude Windeler Kimbel died in April, 1982.

The estate then offered for sale the plantation, consisting at this time of 1,167 acres. That autumn a mobile home park owner took an option on the property. Uneasy over the nature of possible development here, a group of forty-four investors known as Wachesaw Partners purchased that option in March, 1983. Wachesaw Partners promptly hired a team of nationally recognized consultants to design a master plan for the property's development as a private country club community. The investors said their objective was "to preserve the natural beauty of Wachesaw plantation and to offer a degree of privacy, standard of architecture, quality of golfing experience and attractiveness of life style unmatched anywhere north of Charleston."

A conceptual plan along such lines was ready when the actual deed transferring the property to Wachesaw Partners was signed in November, 1983. The price was $4,187,000.

In August, 1984, this group sold the plantation and the development plan to Olan Mills, the well-known photography executive, and his brother C.J. Mills, both of Chattanooga, Tennessee, for $5,000,000. The Wachesaw Partners still retained a minority interest as limited partners.

Since then, property development has basically followed the original conceptual plan. The complex includes a golf course, tennis courts, swimming pool, a clubhouse restaurant and a mixture of single family dwellings and condominiums. A traditional Lowcountry cottage architecture has been used throughout the property. The clubhouse conveys the same idea but is on a grander scale, resembling more nearly the typical Lowcountry plantation house. The Kimbel house, constructed in 1940, was considered too modernistic by the developers for the traditional atmosphere they wished to establish and so was dismantled about 1985. It has been replaced by condominiums now overlooking the Waccamaw River.

During the last few years, the developers have had James L. Michie of the South Carolina Institute of Archeology and Anthropology at the University of South Carolina do extensive archeological exploration here. This effort has resulted in locating old foundations and in finding various relics now exhibited in the clubhouse.

The grounds surrounding the old house site here have always been noted for the large, beautiful live oak trees and the lovely azaleas and camellias beneath them. These have been incorporated well into the new multifamily plan.

Thus has Wachesaw found its place in a new age.

BLACK RIVER

Windsor

NAMED FOR ILLUSTRIOUS WINDSOR CASTLE ON THE banks of the winding Thames River, for centuries the noble residence of English sovereigns, Windsor plantation also lies on the bend of a river. The name Windsor, which means "winding shore," is appropriate for this plantation on the twisting, curling Black River running its course toward Winyah Bay. From Windsor house, a clear, sweeping view extends over the river to old rice fields beyond, and with giant live oaks all about the plantation provides a superb scene of low-country beauty.

Windsor's history begins in the first years of Georgetown County, when the plantation was made up from parts of four different tracts which were combined and then divided to form Windsor and Wedgefield. On May 24, 1734, Robert Johnston, governor of the colonial province of Carolina, granted to William Screven two tracts of land, each containing 200 acres, originally laid out for his father, Samuel Screven, and granted to William in pursuance of his father's will. William Screven and his wife Sarah, then turned over the two pieces of property in November, 1749, to William Poole, who died intestate about 1750. Joseph Poole, William's only son to reach twenty-one, came into possession of both parcels.

A third tract going into formation of Windsor was granted by Thomas Broughton, lieutenant-governor of Carolina, to John Green on June 4, 1735. Listed at 197½ acres, the property later was found to contain 178½ acres more, and these were added to the originally stated 197½. John Green deeded this land to his son John, and made its further disposal subject to the approval of young John's brother William, a co-inheritor of their father's estate. The two brothers in November, 1750, evidently passed the property to a younger brother of Joseph Poole, William Poole the younger, who like his father died about 1750 also without a will. Joseph Poole then became owner of this tract too.[1]

Captain Anthony White, on August 12, 1737, was granted by Lieutenant-Governor Broughton, the fourth tract, consisting of 380 acres. By his will dated February 7, 1756, Captain White gave all his estate to his executors to be sold. In December, 1756, Joseph Dubourdieu, sole executor of Anthony White, let the property go to Joseph Poole, who proceeded to convey all four tracts to John Waties in 1757. John Waties, in his will dated July 5, 1760, directed his executors, Thomas Waties and Andrew Johnston, son and son-in-law, respectively, to sell all of his estate. They disposed of the property in 1762 by dividing it into two portions. One part, including Waties' home, which has been known through the years as Windsor,[2] went to Paul Trapier. The remaining 472 acres went to Samuel Wragg and formed Wedgefield plantation.[3]

From a Berkeley County family of French descent, Paul Trapier was named for his father, with whom began a series of Paul Trapiers in South Carolina. The second Paul Trapier, who bought Windsor from the Waties estate, left the plantation to Paul Trapier III, and he in turn left it on his untimely death in 1778 at twenty-nine to Paul Trapier IV. Paul the fourth lived at Windsor until about 1812, and after suffering financial reverses sold the place to his brother Benjamin, who held it for many years. Benjamin was the father of Brigadier-General James H. Trapier of the Confederate army.[4]

During the War Between the States, the beautiful Windsor house, which is supposed to have been built in 1802 by the fourth Paul Trapier, narrowly escaped burning. There is a tradition that Miss Hannah Trapier, an old lady in her eighties at the time, saved it. Ill in bed when a young Federal officer with a troop of men was sent from Georgetown to burn the house, she refused to leave her bed. Playing skillfully on the officer's sympathies, she told him that she was an old lady and that either way—getting out of bed or staying—she would perish. She upbraided him for his mission of destruction and asked how he would feel were the situation reversed and his mother in such critical circumstances. The officer, flushed with shame, left with his soldiers, and Windsor remained unharmed.[5]

After the war Benjamin Trapier's heirs sold Windsor to the Waldo family, originally from South Carolina but then living in New York.[6] The deed was taken out in the name of Sarah C. H. Waldo in 1870. The Waldos held the plantation only a few years, and on January 30, 1885, the property was sold to Charles Petigru Allston, son of Governor Robert F. W. Allston.[7]

WINDSOR

(Remodeled house not in previous printings)

Born in 1848, Charles Petigru Allston was too young for military service until the last years of the war.[8] His sister, Elizabeth Allston Pringle, wrote of his return at the end of the fighting, when the ragged, starved soldiers of the Confederacy were making their long journeys home: "My dear brother Charley finally came, a ghost of his former self, shoulders bowed down by marching with his heavy knapsack. He looked ill and changed, we were not surprised when we found he had typhoid fever. He had been taken in and kindly nursed by friends on his way home, but he was a pitiful sight."[9]

After graduating from the College of Charleston in 1869, Charles began planting rice and developed eventually into one of the county's most successful post-war planters. His family lived at Windsor for a number of years, and his three children, Susan Lowndes, John Earl, and Richard Lowndes, grew up there. At Charles Allston's death in 1922, the estate passed to his widow, the former Emily Rutledge Lowndes, daughter of Richard Henry Lowndes of Cat Island, and to the three children. Windsor was sold by the family in 1929 to Paul D. Mills of Philadelphia.[10]

A retired investment broker, Paul Mills was the son of General Samuel Miles Mills, one-time Commandant of the United States Military Academy. Paul Mills served in the Spanish-American War, World War I and also at the Versailles Peace Conference. After becoming vice-president of the Philadelphia Electric Company, he resigned to become a broker. Following his death in 1954[11] his family inherited the place and finally sold Windsor plantation on June 27, 1958 to Dr. and Mrs. Harry C. Tiller of Georgetown, who now make it their year-round home.

While wings were being added to the old Windsor home in the spring of 1931, the house burned.[12] Thought to be the third plantation house on the site, it was handsome in both interior and exterior architecture. The lovely hallway, adorned with hand paneling under the curving stair, had ten doorways leading into other rooms. Doors throughout the house had carved ivory handles; and those opening into the drawing room upstairs were especially elaborate, with grooved columns and entablature. Under the house was a high basement built of English brick.[13] The present Windsor residence, designed and constructed along different lines, but in its own right a beautiful, stately house, was erected in 1937.[14]

Wedgefield

LOCATED JUST NORTH OF WINDSOR, WEDGEFIELD HAS much in common with its southern neighbor. The two plantations originated from the same land holding, and today the main residences are strikingly similar in architecture.

The history of Wedgefield as a separate plantation begins in 1762, when Samuel Wragg secured 472 acres of the property formerly belonging to John Waties.[15] Through his will Samuel Wragg left Wedgefield, on his death in 1787, to his sons Joseph and Samuel, both of whom at the age of twenty-one were made executors of their father's estate. The younger Samuel in later years acquired his brother's interests in the plantation. As an old man he sold it in 1840 to Dr. Francis S. Parker of nearby Mansfield plantation. Included in the transfer were the main buildings, mill, and 399 acres.[16]

In 1841 Dr. Parker traded Wedgefield to his father-in-law, the Rev. M. H. Lance, for Greenwich plantation and a sum of money.[17] Mr. Lance lived at Wedgefield for some years, mortgaging it in 1869 to John F. Pyatt, who assigned the foreclosed mortgage in 1875 to Joanna Gaillard. The plantation was eventually conveyed in 1878 to H. M. Haig, who with W. W. H. Holmes had been in actual, continuous, adverse possession of it for over ten years. This adverse possession was sufficient to gain title. Dr. Haig, who was from Charleston, then sold the estate to William M. Hazzard, Elliott W. Hazzard, and John P. Hazzard, all brothers, in 1888. The Hazzards, in turn, transferred the property to their nephew, William W. H. Holmes of Charleston in 1891.

In 1909 Wedgefield was conveyed to Frederick Wentworth Ford of Georgetown.[18] After living there for many years, the Fords sold the plantation in 1920 to James J. Scurry of Georgetown and to E. F. Prosser of Johnsonville, South Carolina.[19] They proceeded to sell in 1922 to the Isaac T. Washbournes of Ossining, New York,[20] who two years later conveyed Wedgefield to Robert S. Mebane and Robert S. Mebane, Jr., of Great Falls, South Carolina.[21] The property changed

WEDGEFIELD

(Natural brick exterior of house not shown in previous printings)

hands again in 1932 when the Mebanes sold to Walter Winchester Keith, a resident of Baltimore.[22] Robert Goelet of Newport, Rhode Island, purchased this old, often-sold plantation in 1935[23] and tore down the Wedgefield house to erect in its place the present handsome dwelling so very similar to the residence at Windsor.[24]

In 1946 Ingram-Dargan Lumber Company of Conway bought the tract[25] and deeded it to Stewart Clare of Georgetown the following year.[26] Clare then sold Wedgefield in December, 1950, to the present owner, John P. Hazzard III, member of an old Georgetown family.[27] The Hazzards spend part of each winter here and reside in Chapel Hill, North Carolina, during the remainder of the year.

Mansfield

THOUGH NOTED ESPECIALLY FOR ITS MAGNIFICENT LIVE-oak avenue, Mansfield has many remarkably preserved and restored buildings that belong to the age of slavery, a fact which adds historic value to the plantation. The late Colonel Robert L. Montgomery of Ardrossan, Pennsylvania, who purchased the estate in 1931, and Mrs. Montgomery, the present owner, were the preservers and restorers of this outstanding old home.

Mansfield plantation is traced to the year 1718, when John Green "had a grant for 500 acres in Craven county [as the Georgetown region was called in the early years] bounding to the northeast on Black River, to southeast on Black River, to the southwest on lands not yet laid out, and to the northwest on lands of said Green, at one shilling per hundred acres."

John Green died in 1750. He had specified in his will that this tract of land be sold and the money divided among his widow Elizabeth, his sons John and Richard, and his daughter Elizabeth. The buyer was James Coachman, a planter of St. James Parish, Goose Creek, who resold it in 1754 to Mrs. Susannah Man, widow of Dr. John Man of Prince Frederick's Parish.

Mrs. Man was the daughter of John and Mary Horry Laroche, descendants of Huguenot families established in the province of Carolina several generations previously. After the death in 1754 of her husband, a property owner of considerable standing in the Georgetown area, Mrs. Man moved to her newly bought plantation on Black River, and since her time the tract has been called Mansfield. Becoming a land owner of consequence herself, she received a grant in 1770 for 400 additional acres of swampland between Black and Peedee rivers and bounding on other lands in her possession.

Dr. and Mrs. Man had two daughters. The elder, Susannah, born in 1745, was married in 1769 to James Cassels, a planter of the Winyah area and of Charleston. Susannah died a year later, leaving an infant son, John, who was entrusted to his grandmother's care at Mansfield.

The husband, who remained loyal to the British government during the Revolution, was charged with having held a commission under the enemy in the conflict, and he was quickly punished with state con-

WINNOWING HOUSE AT MANSFIELD

fiscation of his property. Returning to England about 1782, he took his small son with him. Later Cassels filed a claim in an English court for damages inflicted by the war.

Susannah's sister, Mary, stayed with her mother at Mansfield. In 1772 she purchased land in Georgetown, Mrs. Man following suit with the acquisition of an adjacent lot the next year. The family then moved to Georgetown, but, according to tradition, Mary continued to plant

MANSFIELD

at Mansfield, riding out to the plantation on horseback each day. She became the wife of Archibald Taylor, a Georgetown merchant, in 1785 and had two children, John Man Taylor, born 1786 and Anna Maria, born a year later. Because of ill health, Mrs. Taylor began the family custom—continued until the War Between the States—of spending the summers in a northern climate. She died in Boston in 1801.

Following his graduation from Harvard in 1803, John Man Taylor returned home to the life of a rice planter at Mansfield. In after years he developed poor health. His condition is believed to have been tuberculosis brought on, possibly, by a weakened body, the result of malaria, the deadly disease of this mosquito-infested region. In an effort to improve his health—and probably a depressed state of mind—John decided to try a change of scenery. He went abroad, and while traveling, visited his cousins, the Cassels, of London. Taylor died in 1823 after his return to America. He left Mansfield, through provisions of a will, to his sister Anna Maria, instructing that the place go to her children upon her death. Taylor had never married.

Anna Maria's first marriage was in 1804 to Josias Allston, who died three years later. Her second marriage was in 1815 to the Rev. Maurice H. Lance, a Charlestonian, who was serving then as rector of Prince George Episcopal Church, Winyah, Georgetown. From this union came two daughters to reach adulthood. The younger, Esther Jane Lance, born 1822, married John Harleston Read, of Belle Rive plantation; and Mary, who was born in 1818, became the wife of Dr. Francis S. Parker of Charleston.

Through legal procedure Mrs. Lance turned possession of Mansfield over to her eldest daughter in 1841. Dr. Parker, a graduate of the College of Charleston and first honor graduate of Charleston Medical College, finally gave up medicine to plant rice. The couple made Mansfield their winter home. Upon settling there the Parkers added space to the main house and constructed a small school building close by for their children. They also are supposed to have planted with their own hands many of the live oaks at Mansfield.

Dr. Parker was called upon at a later date to place his signature on the Ordinance of Secession for the state of South Carolina, as the heated controversy between North and South boiled to its angry, tragic climax. Before him, his great-grandfather, John Rutledge, had signed the Constitution of the United States, and two great-granduncles, Arthur Middleton and Edward Rutledge, had signed the Declaration of Independence. Like so many others whose old way of life went with the War Between the States, Dr. Parker was unable to adjust to turbulent, post-war conditions. He died in 1867 at his summer residence near Georgetown—of a broken heart, it was said.

As the war began, the three oldest sons of Dr. Parker joined Confederate forces. The eldest, John, served with Kirkland's Cavalry and was killed by the accidental discharge of a soldier's gun in camp near Dumfries, Virginia, in 1861.

The second son, Frank, became aide-de-camp to General Bragg, rising to the rank of captain and finally to major. After the war he returned to Mansfield. His difficulties with newly freed Negroes and the general state of affairs influenced him to leave, going to Mobile, Alabama, where he later married.

Arthur Middleton Parker, the third son, entered service as a private in Hatches Battalion after his sixteenth birthday in 1861. Later he joined Parker's Battery, commanded by a relative, Captain Edward Parker. In 1869 he married Emma Izard Middleton of Weehaw plantation, and the couple lived for a time at Maurisina, a Lance plantation on Sampit River. They moved to Mansfield in 1885.

The fourth Parker son, James, entered Confederate service with a class of cadets from The Citadel, Charleston, in the last phase of the war. In 1881 he married Penelope Pyatt, daughter of Joseph B. Pyatt of Rosemont plantation and made his home at Waterford plantation on Black River, one of three places he had purchased. The other two were Greenwich and Deer Hall. James died in 1884.

Rutledge Parker, born 1851, lived for some years in Charleston. He returned to Mansfield eventually to manage the place for his mother. After James' death the interests of other heirs to Mansfield were bought by Frank, Arthur, and Rutledge. On Rutledge's death in 1888, Arthur took over operation of the property.[28]

The plantation was sold in 1912 to Charles W. Tuttle of Auburn, New York, who made it a winter home and hunting club.[29] On February 24, 1931, Tuttle conveyed Mansfield, consisting of 760 acres, to Colonel Robert L. Montgomery of Ardrossan, Pennsylvania.[30] After their purchase, Colonel and Mrs. Montgomery energetically went about restoring the buildings. A basement was added to the main house. The small school building in front and the old outdoor kitchen to the rear were converted into guest houses. The kitchen, for such a utilitarian structure, has interesting architecture, with a handsome double stairway adorned by graceful iron railings. Behind the main house and across from the kitchen was built in almost identical architecture a new guest house. A gifted Charleston Negro worker, Thomas Pinckney, who carried out much of the remodeling and restoration of the place, designed and erected the new building, including the curving stairway that duplicates the one of the old kitchen.

STEPS OF OLD KITCHEN, MANSFIELD

As Mansfield was being readied for its new owners, Mrs. Montgomery flew down one hot day to inspect the work. Intrigued by the pattern of the palm leaf fan in her hand, she suggested that the design be followed in a terrace. The ingenius Thomas Pinckney caught the idea immediately, working out the idea with bricks. Strips of turf were left at intervals to form the spokes of the fan. This artfully executed terrace now is a celebrated feature of the plantation.

The Mansfield street, which forms the plantation avenue, still protects beneath the limbs of its huge live oaks the old slave chapel and some of the cabins. Other cabins conforming in style to the original were added by Colonel Montgomery, and in an effort to prevent possible burning of the small cottages, he removed the wooden shingles with which they were sided and replaced them with asbestos shingles especially designed to match the others.

A rice-winnowing house is another old building still standing at the plantation. Elevated from the ground, the winnowing house was used for the sifting of chaff from rice. As the grain was poured through an opening in the floor, the wind would carry off the light chaff while the rice fell into containers.

Also having survived the wear of time is the threshing mill, restored to use when the Colonel in 1939 started planting a few acres of the grain to give to his guests and employees and to use at the plantation. He was forced to give up the operation in 1943 because of war-time labor shortages. Several of these rice fields, which are located to the rear of the house, were banked off some years ago by the Montgomerys and made into a large lake which is now a haven in winter for wild game.[31]

Colonel Robert Montgomery, an outstanding personality among plantation owners of Georgetown County, was born in Radnor, a Philadelphia suburb. He served as a corporal in the Pennsylvania National Guard during the Spanish-American War. Assigned to the signal corps in World War I, he brought his experience as a financier into use by helping to organize the financing and production of the famous Liberty aircraft engine installed in many American planes. He also became a member of the joint Army and Navy Aircraft Board and a member of the General Staff of the Army under General Peyton C. March, then Chief of Staff.[32] An aviation enthusiast, the Colonel was a pilot of both the fixed-wing plane and the rotor plane. He urged American development of the autogiro especially for use in artillery control and trench warfare,[33] an idea which seems to have culminated in the helicopter employed successfully by United States forces in the Korean war.

Colonel Montgomery died in January, 1949, at Mansfield. Mrs. Montgomery still spends each winter at the plantation. In the past Mansfield has been one of the principal estates featured in plantation tours.

Beneventum

FIERY CHRISTOPHER GADSDEN OF REVOLUTIONARY times, the man who was a hero in the hearts of many but only a hothead in the minds of others, was an early owner of Beneventum. Usually associated with Charleston, he is not so well remembered for his ties with the Georgetown region. Yet a large tract of over 1,000 acres, encompassing the present Beneventum, belonged to the famous statesman at his death in 1805.

Gadsden probably acquired the property sometime between 1746, when he returned to Charleston from two years on the high seas, and 1756, when his name is mentioned in a transaction concerning adjoining land.[34] He is known to have prospered during those years and to have gained a respectable holding of property. These were the years, too, before Christopher Gadsden turned to politics. Had he chosen to devote his energies entirely to business, he still would have made a name for himself, but instead he threw himself into the whirl of colonial politics—a plunge that led him eventually to his fame as both statesman and warrior.

Born in 1723, the son of Thomas Gadsden, Charles Town port collector, Christopher received his early education in the colonies and then went to England for more advanced studies. After about two years as a seaman following his school abroad, young Gadsden returned to Charleston in 1746 and became a merchant. By 1761 he owned four stores and considerable property and was enjoying a position of high commercial repute. In 1757 he made his first venture into politics with his election to the General Assembly, where he was to serve for almost thirty years.[35]

As events moved steadily toward the break between the American colonies and Great Britain, Christopher Gadsden, to the horror of more moderate men in Carolina, spoke out for complete separation from Royal control. Not a man of eloquence or of polished oratory, Gadsden spoke in blunt and plain but vigorous language that drove his points home and made many consider him a firebrand.

BENEVENTUM

Gadsden, who had been a representative of the Carolina colony at the 1765 meeting in New York to protest the Stamp Act, was chosen also as a delegate from South Carolina to the First Continental Congress in 1774. He continued to press for American independence, but ironically and to his deep regret, Gadsden was not able to sign the Declaration of Independence, the document he had advocated strongly for ten years. His commission as a colonel of the first regiment of foot of South Carolina necessitated his presence at home.

The South Carolinian could take some satisfaction, however, from the knowledge that his flag picturing the famous coiled snake of thirteen rattles and the words "Don't Tread On Me" became one of the most famous flags of the Revolution. It was presented to Commodore Hopkins as the service flag for the commander-in-chief of the fleet and was hoisted for the first time over an American vessel on December 3, 1775.

After South Carolina's adoption of a constitution in 1778, Gadsden became lieutenant-governor under Rawlins Lowndes and served in the same capacity again upon John Rutledge's election as governor in 1779. He declined the office of governor offered to him in 1782 by the General Assembly. A few years later he voted for ratification of the United States Constitution at a special state convention. In 1790 Gadsden again sat in a convention—this time to frame a state constitution. After earning national rank as a statesman during his long life, this fearless revolutionist died at the age of eighty-one on August 28, 1805, following an accidental fall.[36]

The executors of Gadsden's estate, Philip Gadsden, Thomas Morris and William Drayton, sold the tract containing Beneventum in 1806 to John Julius Pringle, a planter of the Georgetown section, whose widow, the former Mary Izard,[37] later married Joel Roberts Poinsett, for whom the poinsettia is named. With Pringle's purchase of Beneventum, the record of ownership becomes broken and uncertain. The loss of official county records for the period before 1865 makes it impossible in some instances to trace the whole chain of Georgetown property owners. Little seems to be known about Beneventum from the time of its acquisition by Pringle until its sale in 1867 by another owner. In 1825 the property apparently was owned by a Scrivon.[38]

A Charleston merchant, Theodore D. Wagner, held the plantation in 1867 and that year sold it, then some 2,087 acres, to William Miles Hazzard, a Georgetown planter. Hazzard deeded the land and other additional acres to Elliott W. Hazzard in 1883. Elliott's son, William W. Hazzard, as trustee under his father's will, sold the plantation to a real-estate firm in 1910.[39]

Beneventum passed through the hands of two other real-estate companies and finally into the possession of Emerson W. Hitchcock of Auburn, New York, who purchased 600 acres containing the plantation house site in 1916. Holding the plantation until 1931, he in turn sold it to the J. Cornelius Rathbournes of Jefferson Parish, Louisiana. From them Eloise Dudley Drake acquired the place in 1947.[40]

Eloise Drake conveyed Beneventum in 1951 to E. A. Mathiessen, a New York City architect, who sold it in 1953 to Walter W. Posey, Jr., president of the Posey Iron Works at Lancaster, Pennsylvania.[41] From him the plantation was acquired March 4, 1958 by the present owner, Fred B. Lee.

The 220-acre estate still has the handsome Georgian house which the Hazzard family understands is the oldest in this whole section. The two front rooms on the first and second floors are said to have been constructed about 1750 and the other rooms in the main part of the dwelling about 1800. This residence is the original plantation home.

Greenfield

THE STATELY GREENFIELD MANSION LIES IN ONE OF THE most splendidly landscaped settings of the whole region and is a place noted for its camellias and broad lakes. Owned for many years by the late Walker P. Inman, a native of Atlanta, Georgia, and a half-brother of tobacco-fortune heiress, Doris Duke, Greenfield owes its spacious grounds to him. Inman took keen interest in the designing and planting of the magnificent gardens and lakes here, and his efforts turned the plantation into a showplace.

As a rice plantation, the property goes back to 1790 when it was owned by Colonel Thomas Shubrick and called even then by its present name, spelled differently—"Greenefield." A total of 805 acres was included in the tract.[42]

In the years before the War Between the States, Greenfield became closely associated with White House plantation at the fork of the Black and Peedee rivers. Greenfield became the home of John Julius Izard Pringle, and White House was the home of his mother, the former Mary Izard, who after the death of his father, had married in 1833 the noted statesman and diplomat Joel Roberts Poinsett.

Poinsett, a native South Carolinian, served at various times as congressman, first American Minister to Mexico, and as Secretary of War under President Van Buren. Despite his many political achievements, he is perhaps remembered best for the plant he brought from Mexico, known today as the poinsettia. After his marriage to Mrs. Pringle, Poinsett settled at White House. Here he planted a lovely garden; and the plantation, named because of its small white house in the middle of a field, became celebrated for its beautiful grounds.[43]

Poinsett must have won the respect and affection of the Pringles, for his name has been passed down through the generations by the family. John Julius Izard Pringle married Jane Lynch, of New York, and one of their sons, later killed in the War Between the States, was named Poinsett. Another son, Dominic Lynch, gave his son the full name; and Vice-Admiral Joel Roberts Poinsett Pringle showed himself to be entirely worthy of the honor by a career that rivaled that of his step-great-grandfather.

The two other children of John Julius Izard Pringle distinguished themselves by marriage. His daughter Mary married Count Yvan des Franc and went to live in France. Another son, John Julius, married Governor Robert F. W. Allston's daughter Elizabeth, whose book, *A Woman Rice Planter,* was one of the first to tell story of rice planting in South Carolina.[44]

John Julius Izard Pringle eventually came into possession of White House, and in the closing years of the War Between the States, his widow, Jane Lynch Pringle, was operating both Greenfield and White House.[45]

At the outbreak of hostilities, the three Pringle sons were attending Heidelberg University. They immediately left Europe, running the blockade to join Confederate forces. Their parents and their sister Mary remained in Italy until the death of J. J. Izard Pringle in Rome in 1863. Mrs. Pringle and her daughter then returned to America, staying near Philadelphia until they could slip through the lines into Virginia. The two women spent a while there within visiting distance of the Pringle sons and then came on to Georgetown to make their home at White House.[46]

It was at White House that the Pringles had a memorable encounter with an unruly mob of Negroes led by a drunken Federal soldier, who already had pilfered neighboring homes and stirred up the slaves. He came with a large group of followers to the plantation one day, demanding wine and money. Mrs. Pringle, who is described as "bold as a lion and very clever, tall, stout, and of commanding presence, with the face of a man," met them on the piazza and refused to let them enter the house.

According to one account of the incident, she lashed into the soldier with a verbal blasting that threw him completely off balance. She reminded him that she was a northern woman, and she knew the United States Government would not sanction the seizure of her possessions by a drunken mob. She went on to inform him that she was familiar with army regulations against intoxication and added that she had friends in the government, who could see that such an outrage was punished. The angry soldier was on the verge of retorting when Mary began to weep, and his attention was diverted from plunder to her. He tried to calm the girl, even offering her for her consolement

GREENFIELD

stolen goods taken from neighbors. Finally he sought to comfort her with compliments and admiring glances, which so horrified Mary that she fainted full length on the floor. Mrs. Pringle coolly took her limp daughter by the feet and dragged her through the door. On returning to the porch, the stout-hearted lady found that the mob had vanished. Obviously not the type of men familiar with fainting damsels, the drunken plunderers undoubtedly thought they had left a dead woman behind them. Though Mrs. Pringle reproved Mary many times for having failed her in the crisis, it was generally conceded that the young girl's dramatic exit had saved the day.[47]

After the war the D. Lynch Pringles lived at Greenfield,[48] and their son, Joel Roberts Poinsett Pringle, was born at the plantation in 1873. A graduate of the United States Naval Academy, Poinsett Pringle served in World War I as a captain, and in 1926 became chief of staff, commander battleship divisions of the battle fleet. Promoted to vice-admiral in 1926, he became president of the Naval War College the following year. Admiral Pringle died in San Diego, California, in 1932. A destroyer, the USS Pringle, is named in his honor.[49]

Pringle ownership of Greenfield came to an end after many years in the early 1880's. In 1885 Oliver B. Skinner, a successful planter and prominent businessman of Georgetown, acquired the estate. The property was sold in 1919 by his son, J. William Skinner, to Hunter Grover, a Cleveland, Ohio, manufacturer. Grover's interests were assumed in 1923 by the Bank of Georgetown, whose receivers conveyed the plantation in 1935 to the late Walker P. Inman.[50]

Walker Inman was a trustee of the Duke Endowment and of Duke University and had figured prominently in the development of Duke Hospital. In 1947 he was made a director of Duke Power Company. Interested also in aviation and radio, he participated in sending the first trans-Atlantic radio message on short wave.[51]

After his acquisition of Greenfield, Inman built for his year-round home a large brick-and-stucco house, taking the architectural details of the doors and windows from an old Georgian dwelling. The original plantation residence had burned sometime before 1910, and the log structure which replaced it had met the same fate only a few years before Inman's purchase.[52] A nationally circulated home magazine devoted two pages in 1938 to pictures of the present Greenfield house, calling it a "Carolina Classic."[53]

During his early years at Greenfield, Walker Inman planted the formal garden of camellias that today have grown to tree-size. More recently a pathway descending to the lake has been bordered on each side with Pink Perfection camellias. Hundreds of azaleas and other shrubs bank the lakes and winding paths. In blooming season the waters reflect a giant mass of color.

After a lingering illness, Walker P. Inman died at Greenfield in September, 1954, leaving a small son, Walker, Jr., his widow, the former Georgia Polin of Anderson, South Carolina, and a stepdaughter, Susan Hedding Inman. Mrs. Inman eventually married again. She died April 18, 1958. At the present time Greenfield is being held in trust for young Walker Inman, Jr.

Greenfield has been included usually in the various plantation tours.

Keithfield

PICTURESQUE KEITHFIELD, WITH ITS SMALL, QUAINT house surrounded by towering oaks and banks of solidly massed azaleas, is situated on the bluff of Black River. At this spot the channel dips sharply inward, giving a ranging view over the curving stream toward distant rice fields.

Little is known about the early history of Keithfield. The 1825 Mills Atlas places the name of Keith on the approximate spot now occupied by the plantation. This man apparently was John Keith, who is known to have held the land and from whom the property, evidently takes its name. He was dead by 1846; however, this land still seems to have been part of his estate at that time. Before Keith, a William Stitts had possession of the Keithfield tract.[54]

A rice-pounding mill on the plantation was established sometime before the War Between the States and continued operating at least into the 1880's. The mill is mentioned in the business reports of various plantations, but no record has been located to indicate which owner of Keithfield built it.

In 1853 the plantation, which then contained 1,186 acres, was purchased from the estate of Richard O. Anderson by James H. Trapier, son of Benjamin Trapier of Windsor plantation.[55] James emerged from the War Between the States as Brigadier General Trapier, in whose honor a tablet was erected in Prince George Church Winyah, at Georgetown.

Keithfield was acquired from General Trapier's state in 1870 by Mrs. J. J. Izard Pringle and her sons, John Julius and D. Lynch Pringle.[56] From them it went to John P. Hazzard of Georgetown in 1885; and in 1915 the inheritors of Hazzard's estate sold the property, containing 945 acres, to Mrs. Heiman Kaminski, wife of a well-known Georgetown businessman.[57]

Mrs. Kaminski made Keithfield her hobby, going out from her home in Georgetown almost every day to supervise its management. She enlarged and remodeled the small cottage already here, remilling

KEITHFIELD

(House no longer exists)

for the panelling the fine boards from some of the old slave cabins on the plantation. She planted the grounds in azaleas and laid out pathways beneath the huge old oaks shading the house. A bridge was constructed over a low place along the entrance, where water was brought in from the river to form a small pond, not only adding beauty but also serving to irrigate the gardens.[58]

At Mrs. Kaminski's death about 1936, the plantation passed to her son, the late Harold Kaminski. Some 300 acres that included the dwelling and grounds were purchased from the family in 1947 by Dr. Hal B. Holmes of Conway, South Carolina. The remaining acreage was apparently sold as timberland. In the spring of 1955, Dr. Holmes sold Keithfield to W. B. Hazzard, who made it his home.[59] The present owner, L. H. Hutto of Georgetown, purchased the place on March 31, 1959.

Keithfield plantation is one of the spots of great charm and appeal on Black River.

Black River Addenda

Windsor

Windsor was one of the first plantations near Georgetown to be made into a residential community. Today, the main house stands on eight and a quarter acres in the center of the development.

Dr. and Mrs. Harry C. Tiller purchased the property in 1958 and began in 1959 to sell individual house lots on the approximately 300 acres of upland. They reserved nine acres around the Windsor house, where they made their home.

In 1978, the Tillers sold the Windsor house and the accompanying nine acres to Georgetown attorney Kenneth W. Thornton, Jr. While the Thornton family was doing renovation work on the house the following year, the whole center section burned, leaving only the roof and the outer walls. The two wings were not damaged except by smoke. Restoration of the floor, walls and ceiling had taken place by the time the present owners, Mr. and Mrs. Andre Garr of Atlanta, Georgia, bought the place in December of 1986 from Kenneth Thornton. It was also during the Thornton ownership that the attractive portico was added to the river entrance of the house. The portico was added, interestingly enough, to support the expanded attic room being made into a playroom for the Thornton children. The Thorntons were merely trying to make the necessary supports as "aesthetically pleasing" as possible. In the sale to the Garrs, Mr. Thornton retained three-fourths of an acre on the north side of the driveway for a home site.

In January, 1987, final restoration of Windsor began under the guiding hand of Mr. Garr, an industrial engineer who likes to work with houses, too. He obtained a copy of the original plans drawn up in the 1930's by the Charleston architectural firm of Simons, Lapham and Mitchell for Mr. and Mrs. Paul D. Mills, who at that time were coping with the fire that destroyed the old Windsor dwelling. By the end of October, 1987, the house was beautifully restored and furnished. It is now the year-round residence of the Garrs, who have closely adhered to the plantation's spirit in the loving care with which they have brought the handsome Windsor house back to life.

Wedgefield

Like some other plantations in Georgetown County, Wedgefield has been transformed into a country club and residential community. The plantation house constructed here in the 1930's by Robert Goelet, the owner at that time, is now the clubhouse. Stripped of the white paint of the mid-1950's, the house features today its very warm natural brick exterior and, enhanced by recent landscaping, offers a pleasingly attractive appearance. Nearby are the golf course, tennis courts, pool and private residences.

This transformation from plantation to country club began with Wedgefield's sale by Mr. and Mrs. John P. Hazzard, III, on December 8, 1969, to the Equity Development Corporation of North Carolina. Consisting of approximately 600 acres, the tract began to undergo development in 1970. In 1974 Equity incurred financial problems and was forced to cease operations. Everything at Wedgefield stood still for three years. In 1977 the property was purchased by Wedgefield Associates, a Pennsylvania-based group of eight investors. The present general manager, John A. Troup, assumed his responsibilities here in 1979.

In February of 1988, a division of the property was made by the sale of the plantation residence (now called the manor house), golf course, tennis courts and pool operations to George Petrakis and Angelo Bulone of the Hagley area on Waccamaw Neck. The two new owners operate Wedgefield under the business name of WPCC, Incorporated. They have announced plans to upgrade facilities. One business decision already in effect is the leasing of the restaurant in the manor house to John A. Troup. At this time, Wedgefield Associates still retains all rice fields, home sites and the condominium area.

So it is that Wedgefield plantation has followed the local pattern of recent resort development.

Mansfield

Mansfield is one of the few plantations close to Georgetown to remain intact and unchanged through the years. Its preservation has been a remarkable example of good fortune.

Under the appreciative ownership of the Montgomery family

from 1931 until Mrs. Robert Montgomery's death on June 6, 1970, the plantation retained its various outbuildings and historic atmosphere. On October 28, 1970, the 760-acre tract was purchased from Mrs. Montgomery's estate by Wilbur S. Smith of Columbia, South Carolina. Upon his death in 1990, the property passed to his daughter, Sarah Jane Smith Cahalan of Marietta, Pennsylvania. The Smiths took painstaking care of the house, outbuildings and grounds at Mansfield.

Mansfield was placed on the National Register of Historic places in 1977.

Beneventum

For the most part, Beneventum has remained unchanged. Such stability is no small accomplishment considering the ever-increasing residential development all around it. The stately plantation house, which is the eighteenth century original, remains the same and is still surrounded by lovely grounds of live oak trees and blooming plants.

The present owners Mr. and Mrs. Fred B. Lee purchased the 220-acre property from Walter W. Posey, Jr., in 1958. Since then, the Lees have made Beneventum their home and have been faithful in preserving it. The only change here is the reduction in total acreage to 200 acres. Twenty acres were given to their children. This is another plantation fortunately still able to show something of the past.

Greenfield

Greenfield has remained basically the same in recent years. It is still, therefore, a lovely place. Walker Inman, Jr., who inherited the plantation in 1958 when he was a child, still owns it and spends part of each year here.

Keithfield

The destruction by fire in the late 1950's of the charming Keithfield house is the most notable change here in recent years. It has not been rebuilt.

The main house no longer existing when he purchased the plantation in the spring of 1959, Dr. Louis H. Hutto of Georgetown made the old outdoor kitchen into a cottage. He and Mrs. Hutto stayed here when at Keithfield. A guest cottage was made from the remains of the old house. Of the two remaining slave cabins on the place, one remains the same, and the other has been converted into a barn.

Dr. Hutto died on June 28, 1986, leaving Keithfield to his grandchildren. Having a lifetime use of the place are the three Hutto children, Louis H. (Bubby) Hutto, Jr., Ronnie Hutto and Doris H. Wehmeyer.

Though the quaint old house is gone, the setting is still impressive, offering a panoramic view of Black River and former rice fields.

PEEDEE RIVER

Nightingale Hall

IN TIMES PAST NIGHTINGALE HALL WAS MORE NOTED FOR its growing of superior rice than for its musical name. Considered valuable rice property, its fields are located at a pitch of tide where the probability of freshets from above and salt from below is slim.[1] During the years of rice planting, these fields abounded in the golden grain.

In 1769 Nightingale seems to have been part of an enormous tract that contained according to one survey some 8,801 acres of highland.[2] To a rice planter, however, vast acreages of upland meant little; the value of woodland was nothing compared to that of the swamp areas along the rivers.

Robert F. W. Allston of Chicora Wood purchased the plantation in 1846 from the trustees of George Austin Moultrie,[3] possibly the Moultrie who owned the property in 1825.[4] The record is difficult to trace for the years before Allston's ownership. In 1847 Robert Allston added an adjoining plantation, Waterford, a tract of 363 acres situated on the peninsula between Peedee and Waccamaw rivers, to Nightingale and planted the two places as a unit.[5] At Governor Allston's death in 1864, each of his five children was left a plantation and 100 slaves, but Nightingale Hall was reserved to be sold for the benefit of the heirs. Because of its excellent rice land, this property was regarded as the most marketable of the family holdings.[6]

Nightingale, though possessing fine land, had its share of troubles. The slaves here became unruly at various times during the War Between the States and at the end of the conflict, openly dangerous. On Mrs. Allston's return from Crowley Hill high up the Peedee, where the family had taken refuge from the threat of coastal raids, the Negroes at Nightingale Hall had locked the overseer in the house, stationing a guard outside with a shotgun. They threatened him with death should he dare try to escape. The overseer, relieved that his fate was to be no worse, assured them he had no thought of leaving. He then settled himself near a window to keep account of the goods being

NIGHTINGALE HALL

stolen from the outbuildings, and thus in good spirits passed the days of his imprisonment.

When the Governor's widow and her daughter Elizabeth drove into the yard, the Negroes quickly gathered. Mrs. Allston, after easing into general conversation with the colored foreman, asked to be given the keys to the barn. Replying that "de officer" had left them with him, the foreman refused. When Mrs. Allston showed him a written order for return of the keys, the crowd seemed to become more sullen; and a young Negro, aiming a stone at her from time to time, edged up to the carriage. Realizing the critical situation, Mrs. Allston calmly stepped from her carriage into the midst of the throng. She walked among the crowd, talking to each person by name and inquiring about the different families. Gradually the tension relaxed. The foreman handed over the keys, and the two women departed.[7]

In 1869 Nightingale was purchased from Governor Allston's estate by Edward N. Thurston of Charleston as trustee,[8] and he sold in 1891 both Nightingale Hall and Waterford to John Richardson of Georgetown.[9] Richardson in turn conveyed the 1,219 acres of Nightingale to Jacquelin Smith Holliday of Indianapolis, Indiana, in 1917.[10] To this purchase the new owner added that year the two adjoining places—Clermont, 857 acres, and Ingleside, 465 acres.[11]

Charles N. Ingram and E. Ervin Dargan, prominent businessmen of Conway, acquired the three plantations from the trustees of Jacquelin S. Holliday in 1948.[12] After a twelve-year ownership, the Nightingale Hall property was conveyed by the Dargan connection on June 1, 1960 to the current holder, Metropolitan Properties, Incorporated.

Milton

THE PROMINENT IZARD FAMILY OF SOUTH CAROLINA held Milton plantation through four generations before finally giving it up in 1880. Their ownership covered almost the entire nineteenth century. Ralph Izard, Jr., who was born in the 1750's and who died sometime between 1800 and 1810, one of many Ralphs in the family, appears to have been the first of his line to own Milton, along with three other Peedee places, Weymouth, Hickory Hill, and White House.

Born at Fair Spring, St. George, this particular Ralph Izard completed his education in England and after returning to South Carolina was elected to the Provincial Congress in 1775 from St. John's, Colleton. He later served in the General Assembly from St. George and St. Andrews and in 1788 sat in the convention on adoption of the Federal Constitution. Though Izard held sizable acreages on Peedee River, he apparently never lived on any part of these lands. His last years were spent in Charleston.

After the death of Ralph Izard, Jr., his son, Ralph Stead Izard, born in 1783, inherited the estate, which still included the Peedee rice plantations. However, he lived to enjoy his possessions only a short time, for he died in 1816. His sister was the Mary Izard who married John Julius Pringle and Joel Roberts Poinsett. Ralph Stead was the only son among seven children. His widow, the former Esther Middleton, daughter of Colonel Thomas Middleton, survived him by three years.

Milton plantation next belonged to Ralph Stead Izard, Jr., who was an infant of one year when his father died. Following a European education and several years of foreign travel, Ralph settled down to the management of his plantations, Weymouth and Milton, spending the winters in South Carolina and the summers at Newport, Rhode Island. In 1839 he married a cousin, Rosetta Pinckney, daughter of Colonel Thomas Pinckney, Jr. Ralph Stead Izard, Jr., died at Newport in 1858.[13]

MILTON

Of Ralph and Rosetta Izard's six children, a daughter, Josephine Louisa, came to own Milton. Born in 1850, Josephine married William DeCourcy May of Baltimore, Maryland, in 1876 and in 1880 sold the plantation to Charles Petigru Allston, son of Governor R. F. W. Allston. Milton contained at the time of this transaction some 511 acres.[14] Five years later Charles Allston purchased Windsor on Black River, where he made his home for many years.

Milton passed from Allston in 1901 to J. B. Steele of Georgetown, who then conveyed the tract in 1911 to James Louis LaBruce, Jonathan I. Hazard, and G. Albert Lohr. Eventually the land came into the sole possession of Jonathan Hazard, and the heirs to his estate conveyed the property to Henry H. Groves of Gastonia, North Carolina, in 1941.[15]

The present Milton house, a small structure, was built originally during the Groves ownership as a camp for use during hunting season. It was enlarged and remodeled to the present proportions by A. M. Quattlebaum. A driveway recently planted with young oaks forms the entrance leading to the house situated on a point of land edging the marsh.

Trustees of the Groves family sold the plantation in 1953 to E. Ervin Dargan of Conway, who immediately transferred 411 of the 511 acres in the tract to Alexander M. Quattlebaum of Florence,[16] who transferred the holding back to Dargan in September, 1955. It was acquired the same month by Herman A. Moore of Charlotte, North Carolina. The turn-over of ownerships continued with the deeding of Milton Hall in May, 1960 to The Hagley Corporation and on to The Jericho Corporation in February, 1961. Here, for the time-being, the property rests.

Dirleton

ONE OF THE BETTER KNOWN PLANTATIONS OF GEORGE-town County, Dirleton on the Peedee received its name from the Heriot ancestral seat, Dirleton Castle in Scotland, where the ruins of this ancient stone structure still may be seen.[17] The plantation goes back to the lifetime of Dr. Edward Thomas Heriot, son of William Heriot, who came to America from Scotland to join his brother, Colonel Robert Heriot. An eminent American soldier of the Revolution, Robert owned a large part of Hobcaw Barony at one time.

Dr. E. T. Heriot acquired the land now known as Dirleton in 1854. It was then called Richfield. However, he lived at another plantation, Mont Arena, on Sandy Island between the Peedee and Waccamaw rivers and owned also Northampton on Sampit River. Upon Dr. Heriot's death, Dirleton passed to his eldest son, Robert Stark Heriot. It was with him that Dr. James Ritchie Sparkman exchanged his own place, Birdfield, for Dirleton in 1859. Dr. Sparkman had married in 1845 Robert's sister, Mary Elizabeth Heriot. Upon the Doctor's move to Dirleton, the long Sparkman reign here began.[18]

James Sparkman (born in 1815) lived to the advanced age of eighty-two and at his death in 1897 had established himself very prominently in the life of the coastal region. He was held in high regard by fellow citizens, and plantation owners of the entire area looked to him for medical service for both families and slaves. He began his medical practice in the Peedee section in 1836. Most of his time up until the War Between the States was spent as the regular medical attendant and adviser to more than twenty-five plantations for some twenty miles around, holding a population of about 3,000 Negroes in all.[19]

Dr. Sparkman built with plantation labor the spacious house at Dirleton, considered even today one of the sturdiest on the river.[20] Its three floors contain twenty-three rooms, each one of which on the main level has a marble fireplace. Typifying the piazza-style dwelling of many low-country plantations, the Dirleton house was built in the late 1850's, but final touches were added after the War Between the States.[21]

Food for both humans and animals became so scarce during the last months of war that Dr. Sparkman was unable to continue his practice over the large Peedee plantation area and took his family up to Clarendon District, where conditions, were at least a little better than at Dirleton. There the family stayed for a while, even after Lee's surrender. Left completely bankrupt, Dr. Sparkman could not re-establish affairs at his plantation immediately, and the continued scarcity of provisions made his return difficult. In the family's absence Dirleton fell victim to the whims of the former slaves, and often their notions resulted in drunken parties.[22] However, the main house survived to become once again the Sparkman winter home.

In 1858 Dr. Sparkman wrote an invaluable report to the state government describing slave life on the average rice plantation. Using as an illustration his own plantation of one hundred laborers, the Doctor described work, food supply, clothing, housing, medical attention, and religious instruction among slaves. Not only does the paper discuss the complex organization needed for the successful operation of a plantation, it also reveals the fair, humane treatment of the many souls that were his responsibility. Dr. Sparkman allowed slaves on his place to earn money of their own through the sale of livestock and home-manufactured items. He himself was their best customer.[23]

Surprisingly enough, many of the most colorful events in Dirleton's history came not before the war but in the 1890's, the last years of Sparkman ownership. During those years the dashing Manigault Light Dragoons, composed of Georgetown County men, held drills and contests of swordsmanship at the plantation. The thirty gray-uniformed horsemen took part in tilting tournaments, while their ladies cheered from the sidelines. The best swordsman was given the privilege of crowning a girl of his choice "queen of love and beauty"—an honor won each time by the company captain, James Louis LaBruce of Arundel plantation. Since meetings of the Manigault Light Dragoons usually lasted all day, the spectators came with picnic lunches and settled down comfortably to watch proceedings. Officially part of the South Carolina Militia, the Dragoons thrived on feminine interest and seem to have fulfilled more of a social than military function.[24]

DIRLETON

At Dr. Sparkman's death on November 27, 1897, Dirleton passed to his widow, Mary Elizabeth Heriot Sparkman, who died in 1912, leaving the plantation to her children. Later, some of the heirs, including a daughter, Mrs. Allen Hamby, a grandson, George E. T. Sparkman, and the surviving children of Jane Sparkman Ward sold their interests in the place to Hattie Buck Sparkman, the wife of Dr. William Sparkman, one of Dr. Sparkman's sons.[25] The 1,094 acres of Dirleton were then sold in 1917 by Edward Heriot Sparkman, eldest son of Dr. J. R. Sparkman (executor of his father's will and administrator of the estate) for himself and the other heirs, Sullivan Thorne Sparkman, J. R. Sparkman, Jr., William Ervine Sparkman and Hattie Buck Sparkman, to Philip R. Lachicotte, son of St. Julian M. Lachicotte of Waverly on the Waccamaw, and to L. J. Upton and T. R. Upton of Norfolk, Virginia.[26]

The plantation eventually came into the hands of John Upton of Norfolk as sole owner, and he returned it to the Sparkman line by sale to Louis L. Hamby in 1932.[27] After holding the plantation for over a decade, Hamby conveyed Dirleton in 1943 to the Charles N. Ingram–E. Ervin Dargan business connection of Conway. The tract was reduced to 624 acres at this transfer, some 470 acres of timberland having been purchased a year previously by the International Paper Company of Georgetown.[28]

A noted publisher of books on firearms, Thomas G. Samworth of New Castle County, Delaware, acquired the plantation in 1945 from Ingram and Dargan and now makes it his year-round home.[29] In 1961 he deeded Dirleton and 600 acres of land to the state of South Carolina as a gift with the stipulation that he and Mrs. Samworth retain life tenure of the plantation house. The South Carolina Wildlife Resources Commission plans extensive water control work here to improve the land for waterfowl. The plantation has always been exceptionally good duck hunting ground.

Dirleton has been included often in plantation tours of the Peedee region.

Arundel

NATURAL BEAUTY AND HISTORICAL ARCHITECTURE ARE the distinguishing marks of Arundel. The plantation has not only a live-oak avenue and a magnificent grove of oaks covering acres of ground but also many ante bellum buildings which have been kept intact throughout the years.

All of the outbuildings at Arundel are constructed in a Gothic style, modified to suit the needs of a South Carolina rice plantation. Two of the fifty slave cabins that once lined the street remain. In addition there are the overseer's cottage, a quaint structure estimated to be over one hundred and fifty years old; the smokehouse, another interesting piece of carpentry; the ration house with its barred windows; and a small guard house located almost in front of the ration house—erected originally to prevent unauthorized appropriation of food supplies. Most of the little Negro cabins have deteriorated over the passing years, the expense of maintaining them having outweighed by far the small need they filled in modern plantation operations. However, those still standing, together with the other buildings, offer a superb "panorama of slavery," as the situation has been described.

Until the spring of 1955, the plantation was the home of Mrs. Herman Carraway, formerly Mrs. James Louis LaBruce. Following Mr. LaBruce's death in 1915, Arundel remained in his family; the plantation's recent sale to the present owner, Alexander M. Quattlebaum, marked the end of the LaBruce regime at this lovely estate.

The LaBruce family is believed to have been the first occupants of the present main house, despite the fact that the dwelling was erected years before Arundel came into their possession. Hanging in the hall is an 1827 plat of the property which shows the original house to have been located some distance north of the present one. The first residence burned sometime between 1827 and 1841, the approximate year that work began on the two-story home now used. The house was not completed fully until James Louis LaBruce established his family at Arundel in the late 1890's.

ARUNDEL

Generally accepted tradition says that this Peedee plantation received its name from Arundel Castle on the Arun River in Sussex, England, a sketch of which appears in a 1799 copy of *European Magazine* in Mrs. Carraway's possession. Also at the plantation is a miniature photograph of a large oil painting, now on exhibit in London, of the second Earl of Arundel.

Dr. William Allston is known to have owned the Arundel tract for a time during the first half of the nineteenth century. A memorandum prepared by Mrs. Carraway states that Dr. Allston acquired the property from John Julius Pringle in January, 1806; and this information, apparently based on papers at the plantation, is quite possibly accurate, although official records for the conveyance cannot now be discovered. Dr. Allston sold Arundel in 1841 to Frederick Shaffer, a northerner, who built the present main house, although neither Shaffer nor his family ever lived there.[30]

In 1860 Shaffer's heirs sold the estate to Charles Alston, of Bellefield plantation on the Waccamaw, who willed it to his son, Charles Pringle Alston. Louis Claude Lachicotte, son of Philip R. Lachicotte of Waverly, then purchased Arundel from Charles Pringle Alston on December 1, 1897. Part-owner with his brother, Francis W. Lachicotte, and with James Louis LaBruce, son of Joshua W. LaBruce, in a rice-planting firm known as the Guendalos Rice Company, Louis Claude sold the plantation in 1899 to the company.[31] LaBruce eventually bought out all interests of the Lachicotte brothers and became sole owner of Arundel in 1904. The LaBruce family actually had been living here earlier. After purchasing Guendalos Rice Company, Louis LaBruce held besides Arundel, the plantations of Springfield, Enfield, Exchange, Guendalos, and Birdfield.

The large grove of live oaks to the right of Arundel avenue is supposed to have been planted by the LaBruces soon after they came here to live. Hundreds of azaleas set out in those years now line the circular driveway and paths at the house. Here on the bluffs of the famous Peedee each spring, these gorgeous shrubs, protected by tremendous live oaks towering proudly over the gardens, offer a scene of breathtaking color and beauty.

In 1894 James Louis LaBruce married Georgie Alberta Stokes,[32] daughter of the Rev. G. W. Stokes of Vaux Hall, Waccamaw, and had two sons from this marriage to grow to maturity—the late James

Louis, Jr., and the late George Albert LaBruce, who jointly inherited Exchange plantation farther up the Peedee.

During the few years that the Manigault Light Dragoons, a military company of Georgetown county men, existed in the 1890's, Louis

SLAVE CABIN AT ARUNDEL

LaBruce was the company captain and consistent winner of every tilting tournament ever staged by the organization. He often invited the Dragoons to Arundel for winter meetings that lasted usually all day and featured drilling and contests of swordsmanship.[33]

Mrs. LaBruce died in 1901. The following year, James Louis married Esther Richardson, member of a prominent family of the state and a descendant of Davison McDowell, one-time owner of Exchange. The four children from this second marriage are Allard, Leger, Lawrence, and Esther. James Louis LaBruce died in 1915.[34] His widow continued to live at Arundel and later married Herman Carraway, of Georgetown, who died in 1943.

Arundel was the home of Mrs. Carraway until its purchase in 1955 by E. Ervin Dargan, who transferred it in May of the same year to Alexander M. Quattlebaum, of Florence, S. C., who plans eventually to live here throughout the year.

The plantation has been a leading place of interest in the various plantation tours.

LIVE OAKS AT ARUNDEL

Prince Frederick's Church, Peedee

FROM ITS VERY BEGINNING, THE EPISCOPAL PARISH OF Prince Frederick's, Peedee, has had a life filled with misfortune and tragedy. A slow death now grips the old church, as the handsome building, abandoned some years ago, is being given up to decay.

Comparable in architecture to the beautiful churches of the English countryside, this building was constructed in 1859 but not fully completed until after the War Between the States. For some years following the end of rice planting and the resultant exodus from plantations, services were alternated between this main church on the river road and the summer chapel at Plantersville; and by 1934 services were being conducted altogether in the more accessibly located chapel.

Prince Frederick's, originally a part of Prince George Parish, was established as a separate parish on April 9, 1734, by act of the General Assembly of South Carolina. It was named in honor of Frederick, Prince of Wales, son of King George II. Elizabeth Allston Pringle, who published the records of this church, comments that the unsaintly names of Prince George and Prince Frederick's churches appear to have influenced the early histories of both. Prince Frederick's Parish first took as its main seat the wooden structure high up Black River at the point now called Brown's Ferry. The building had been erected about 1726 to serve as Prince George Church, but time proved the location undesirable. When the parish of Prince Frederick's was formed, this building was transferred to the new group, and plans were made for the construction of the present Prince George, Winyah. Church at Georgetown. The new territory of Prince Frederick's included the old Black River site; and here the parish functioned until an outstanding former rector, the Rev. Hugh Fraser, donated in 1835 another, more convenient site on the Peedee River, where the church subsequently was established.[35]

In 1755 began one of the most tragic chapters of British rule on the American continent—one that had direct connection with Prince Frederick's Parish. The Acadians were taken from their homes in Nova

Scotia and distributed among the various colonies, and a group was dumped on the unhappy authorities at Charles Town. The officials decided to spread the Acadians among the settlements and sent to the parish of Prince Frederick's one of the largest groups to go to a rural section. On August 10, 1756, some thirty-two Acadians were distributed indifferently among local settlers. Eight of the newcomers were dead within three months—it is said, from malaria or heartbreak.[36]

After its establishment as a parish, Prince Frederick's was apparently served by Thomas Morritt, first minister of Prince George, and John Fordyce, who succeeded him in 1736. Mr. Fordyce was outstanding in coastal missionary work as far up as the Cape Fear River. After his departure the vestry had a difficult time with clergymen. They were described as "good, bad and indifferent—the second class apparently prepondering" on into the Revolution.

Hugh Fraser, a native of Scotland, became rector in 1783 and served until 1810. He kept the church on its feet during troublous times following the Revolution, when the American Episcopal church had difficulty in getting priests ordained and the shortage of clergy was crucial. Mr. Fraser resigned in 1810 and in 1812 accepted the rectorate of All Saints, Waccamaw, where he remained until 1817. He returned to Prince Frederick's twenty years later, to serve again from 1837 until the following year. While at All Saints he had conducted occasional services for the Prince Frederick's members.

The leadership and personal generosity of the Scottish clergyman were directly responsible for erection of the first church on the site of the present building. In 1835 he donated the land and gave with it a right-of-way to Peedee River only a short distance away. The site was surveyed by Robert F. W. Allston, and a building committee was appointed, consisting of Allston, Francis Weston, and Davison McDowell. With consecration of the new building in 1837 by Bishop Nathaniel Bowen, Prince Frederick's began a period of prosperity and development that continued until the War Between the States.

From 1846 through 1862, Joseph Hunter was rector. His success was such that in 1857 he reported the number of communicants to be nearly double that of the preceding year. At a meeting the same year, need for a larger building was discussed, and in 1859 work began on the new Prince Frederick's Church, which was to be located on the site of the old structure. Laying of the cornerstone and an address by

PRINCE FREDERICK'S CHURCH, PEEDEE

(Building no longer exists)

Governor R. F. W. Allston, senior warden, took place on December 17, 1859. Work continued into 1860 but soon was interrupted. Materials from Europe were lost in the Federal blockade of southern ports, and the contractor, Gunn, was killed during construction. The builder's name lives on, however, for the church is often called "Gunn Church."

The partly finished building became so damaged during and after the war that the whole interior had to be redone. Impoverished members were forced to wait until 1876 when John Earle Allston of Brooklyn, New York, presented a gift of $1,700, which made possible final repairs and completion. Additional gifts came from other sources. J. D. Sparkman of New York had given a marble baptismal font in 1860. Choir stalls, a lock, and bolts and hinges for the doors came from the beautiful chapel at Hagley, a Weston plantation on the Waccamaw, and pews were contributed by All Saints church.

Prince Frederick's Peedee, was consecrated by Bishop Howe on Palm Sunday, March 25, 1877. Taking part in the service were Dr. James R. Sparkman of Dirleton, a long-time officer of the parish, who read the request for consecration; the Rev. Alexander Glennie, rector of Prince George, Winyah, formerly rector at All Saints and during the years from 1862 to 1865, rector of Prince Frederick's; and the Rev. William H. Barnwell, who was then rector of All Saints.

At the opening of new Prince Frederick's, the old building was moved to Plantersville to replace the crudely constructed chapel erected there in 1848 and had fallen into bad condition by 1877. The new chapel was consecrated by Bishop Howe on July 23, 1879, during the ministry of Benjamin Allston, son of Governor Allston.

Benjamin Allston came to Prince Frederick's as rector in 1877, serving until 1890, when he resigned to go to Union, South Carolina. His work on Peedee was outstanding. In addition to his work at Prince Frederick's, he established St. Cyprian's chapel for Negroes and founded St. Peter's mission. Elizabeth Allston Pringle, the rector's sister and a devoted member of the church, later compiled and published a volume of Prince Frederick's records.

Serving the parish next as rectors were LeGrand F. Guerry, Charles E. Cabiness, John E. H. Galbraith, Charles W. Boyd, and Henry D. Bull—all also rectors of All Saints, Waccamaw.[37] After Dr. Bull's resignation in 1953 following twenty-eight years of faithful work, the Peedee congregation secured as their minister, Edward Small, rector of the Episcopal church in Conway, South Carolina.

At the present time, the Rev. Roderick J. Hobart, rector of St. Paul's Church, Conway, serves as minister-in-charge of Prince Frederick's.

Today an air of sadness surrounds old Gunn Church, as its weather-beaten walls and interior fall daily into greater dilapidation. Obviously caving braces show that the building cannot stand many more years, but its massive lines are still impressive. Standing high amid tall pines that reveal at a distance intriguing glances of its steeple, this church of the Old South seems reluctant to sink into the oblivion apparently decreed for it by the modern world.

There is small chance that the building can ever be restored, but, even so, church officials in 1954 began clearing and restoring the churchyard. By the summer of 1955, a sturdy metal fence had been erected across the entrance, and the reclamation of the grounds had been completed. Though it must remain lifeless, Gunn Church lay, for the moment at least, in attended surroundings.

Chicora Wood

CHICORA WOOD, SINCE LONG AGO A DISTINGUISHED name among South Carolina plantations, has been made famous by the pen of Elizabeth Allston Pringle, who brought its early years vividly back to life in *Chronicles of Chicora Wood* and *A Woman Rice Planter.*

Mrs. Duncan Cameron Waddell of Asheville, North Carolina, the present owner, shows her keen appreciation of history and setting in the devoted attention she gives to this Allston home, built before 1819. Chicora Wood's importance as the residence of Governor Robert Francis Withers Allston is paralleled by its importance as the home of his daughter Elizabeth, and as the locale of her two books.

The 915 acres presently contained in Chicora Wood property were spread originally among four separate tracts and part of a fifth. Two of the tracts containing together 520 acres of both swamp and highland were granted in 1732 and 1734 to John Allston, one of the two brothers from Berkeley, who established the family on Waccamaw Neck. John Allston in 1738 bought the third tract of 246 acres from William Waties, to whom it had been granted in 1736. Ten years later John conveyed the first two tracts and 150 acres of the third to his son, John, the land passing from him in 1751 to his daughters, Mrs. Benjamin Young and Mrs. Peter Simons. In 1764 Mrs. Young's share was purchased by Peter Simons, who sold all to Benjamin and Richard Waring in 1771. After acquiring Benjamin's interests, Richard Waring sold the land in 1777 to John Waring. Earlier that year John Waring had secured from Benjamin Walker two adjoining tracts amounting to 252 acres. All of these various lands together made up the future Chicora Wood property. Holding the plantation only two years, John Waring sold out to Alexander Rose and John Torrans, Charleston merchants. Rose then purchased Torrans' portion from his widow in 1785, and on July 19, 1806, Benjamin Allston, Jr., grandson of John Allston, purchased the property from executors of Rose's estate.

In his will made out in 1807, Benjamin willed the Peedee River plantation to his younger sons, Robert and William. The eldest son, Joseph Waties, was left the family home, Waverly, on the Waccamaw. During Mrs. Allston's management of the plantation belonging to her younger sons, William died. His share then was divided between the other brothers. Robert eventually bought out Joseph to become full owner in 1832.[38] At that time Chicora Wood was called Matanzas, for a locality in Cuba,[39] from which, it is believed, slaves were brought to the plantation.[40] Not until 1853 was the name changed to Chicora Wood,[41] a combination that Mrs. Robert Allston liked.[42]

After Benjamin Allston's death in 1809, Mrs. Allston, who was the former Charlotte Ann Allston and her husband's cousin, undertook the task of managing her three sons' property—a burden that wore on her more and more. She frequently voiced her distaste for the task, and indeed a full share of troubles justified her discouraged state of mind. Finally turning Waverly over to Joseph in 1819 as he came of age, she moved to Chicora Wood to concentrate her energies on difficulties there. A dispute with a neighboring planter having arisen over boundaries, she was forced to seek help from her son, Robert, who was serving as an officer in the United States army. Young Allston thus resigned his commission in 1822 to come home and become a planter.[43]

Born in 1801, Robert Francis Withers Allston was the second son of a family of six children.[44] He attended school in Georgetown until he was sixteen, when he entered the United States Military Academy in 1817. He graduated in 1821 and was on duty with the army engineers at Mobile Bay the following year, when his mother's struggles at home brought him back to Georgetown. Robert soon got the boundary issue settled favorably by court action, and on his mother's death in 1824 assumed full management of his plantation.[45]

The lawsuit was to prove of greater significance than anyone realized at the time. Employing for his case the renowned lawyer, James L. Petigru, Allston came to know Petigru's sister, Adele, and in 1832 married her. Robert and Adele Allston made their home at Chicora Wood in the house that stands today.[46] Of their ten children, five lived to maturity: Benjamin, who also went to the United States Military Academy, later becoming a rice planter and eventually entering the Episcopal ministry; Charles Petigru, who bought Windsor on Black River, where he planted rice for a number of years; Adele, who married Arnoldus Vander Horst; Elizabeth Waties, who became the

CHICORA WOOD

chronicler of Chicora Wood and Jane Louise, who married Charles Albert Hill.[47]

An intimate portrait of Robert F. W. Allston's remarkable character may be obtained from Elizabeth Pringle's *Chronicles of Chicora Wood,* and a full account of Allston's business operations is given in J. H. Easterby's *The South Carolina Rice Plantation: As Revealed in the Papers of Robert F. W. Allston.* This sketch can hope to catch only the highlights in the life and career of the distinguished governor and planter.

R. F. W. Allston left the imprint of his strong personality on everything he touched. He was a man of high principles, sound judgment, and great versatility. An overseer once said that Robert Allston's tremendous success as a planter was due not to any special agricultural abilities but to his gift for organization; his plantations were made to function in an orderly, systematic manner. A scientific rice planter, Allston often wrote articles for *De Bow's Review* that were accepted as authoritative. In 1855 a sample of rice grown at Chicora Wood won a silver medal at the Paris Exposition.[48]

The military education Allston had received made it possible for him to fill the office of Surveyor-General of South Carolina, to which he was named in 1823. He was elected to the State House of Representatives from Prince George Parish in 1828 and was re-elected two years later. Defeated for the House in 1832, he became a candidate for the South Carolina Senate in a special election later that year and made it—by one vote. He remained in the Senate for the next twenty-four years, serving as president of that body from 1847 to 1856, when the General Assembly elected him to the governorship. He held the office of governor through December 14, 1858. Basing his political philosophy on the principles of government as expressed by Thomas Jefferson, Governor Allston was a staunch believer in States Rights, and he never seems to have doubted the wisdom of the Southern position on slavery and on distribution of governmental power.[49] From 1841 until 1864, he was a trustee of South Carolina College, now the University of South Carolina.[50]

Allston's prosperity was such that when he died on April 7, 1864, he left a plantation and a hundred slaves to each of his five children.[51] He had come into possession of seven rice plantations during his lifetime and besides his own estates had managed for twenty-three years Waverly plantation, which belonged to his two nephews. Waties

Point, located at the edge of Georgetown, willed by his aunt in 1840 to him as trustee for his sons and nephews, was sold in 1854. Another part of his aunt's estate, Friendfield, on Waccamaw was sold by Allston after one year, enabling him to make other purchases closer to Chicora Wood.

In the 1840's Allston began expanding in earnest. He purchased Exchange, two miles up the Peedee from Chicora, in 1843, sold it three years later, and then re-bought it in 1853. Nightingale Hall was acquired in 1846, and the adjoining Waterford plantation the next year. Rose Bank, which lies just north of Chicora, was obtained in 1851. (Allston called it Ditchford or Ditchfield, rather than Rose Bank as the plantation is known today.) Sometime before 1853 he purchased Breakwater, and in 1858 he bought Guendalos, another Peedee estate. Pipe Down, on Sandy Island, was acquired the following year. Thus R. F. W. Allston came to hold vast properties and slaves—an awesome responsibility that, no doubt, hastened his death during the War Between the States.[52]

Governor Allston was adored by his children and he had intended to do well by them. According to his will, Chicora Wood went to Charles Petigru; Guendalos to Benjamin; Ditchfield to Jane; Exchange to Elizabeth; and Waterford to Adele. Nightingale Hall was to be sold for the benefit of all the heirs.[53] As things turned out, nothing went according to Governor Allston's plans. At the end of the war, there were heavy debts against the estate that could not be paid, and in 1869 the Allston property was sold. Out of the huge land holding, only Chicora Wood, awarded to the widow as her dower, was saved. Here Mrs. Allston lived and planted rice until her death in 1896.[54]

Mrs. Pringle describes the family's return to Georgetown at the end of the war. On going to Chicora Wood to inspect damage inflicted by former slaves, she says:[55]

> We looked at the house; it was a wreck,—the front steps gone, not a door nor shutter left, and not a sash. They had torn out all the mahogany framework around the doors and windows—there were mahogany panels below the windows and above the doors there were panels painted—the mahogany banisters to the staircase going upstairs; everything that could be torn away was gone. The pantry steps being there, we went into the house, went all through, even into the attic. Then the big tank for the supply of the water-works, which was lined with zinc, had been torn to pieces, and the bathroom below entirely torn up. It was a scene of destruction, and papa's study, where he kept all his accounts and papers, as he had done

from the time he began planting as a young man, was almost waist-deep in torn letters and papers.

The Allstons had refugeed during the war at a point high up the Peedee because of the probability of coastal raids, but at the place of refuge, they, like many others, stepped into the path of Sherman's army. With such an experience behind them, the family probably were prepared for what they saw at Chicora Wood.

Elizabeth Waties Allston, after the war, married John Julius Pringle of the family that owned Greenfield and White House plantations. Their marriage was cut short after just six years by the death of John Julius. Young Mrs. Pringle later purchased White House plantation and, following her mother's death, Chicora Wood, where she spent the rest of her life.[56]

Mrs. Pringle planted rice for some years at her two plantations and between 1904 and 1907 wrote articles on her experiences as "a woman rice planter" for the *New York Sun*. Later the articles were collected and published in book form under the title, *A Woman Rice Planter*.[57] Illustrated with Alice R. Huger Smith's quaint sketches that capture so charmingly the spirit of plantation life, the book has come to be regarded as one of the best ever written on the rice-planting period. Unfortunately it has been out of print for many years.

Mrs. Pringle's second book, *Chronicles of Chicora Wood*,[58] is a fascinating account of ante bellum plantation life, of the family's encounter with General Sherman's army, and of life at Chicora Wood after the war. Brought out in a second edition a few years ago, this book, too, is now out of print. The literary contributions of Mrs. Pringle have taken their places as classics on South Carolina's bygone rice age.

With Elizabeth Allston Pringle's death in 1921, Chicora Wood passed out of the Allston family. Her younger sister, Mrs. Hill, and Elias Vanderhorst, executors of Mrs. Pringle's will, sold the plantation in 1926 to the late Duncan Cameron Waddell, Jr., of Asheville, North Carolina.[59] Mr. Waddell, with loving and appreciative hands, fully restored the Chicora mansion, and reclaimed the grounds, including the beautiful grove of live oaks that had been swallowed up by the forest. With the exceptions of the banksia roses, wisteria, and a few azaleas, the present gardens were laid out and planted during his time.[60]

Great care, too, was put into the restoration and preservation of many outbuildings at the plantation. Chicora has one of the few remaining rice-threshing mills in the county—one that is in exceptionally fine condition. Among the other old structures which have been saved by the Waddells from the decay of time are the small schoolhouse once used by the Allston children, the outdoor kitchen, the brick smokehouse, and the barns.

KITCHEN AND PLANTATION BELL AT CHICORA WOOD

Since Mr. Waddell's death, Mrs. Waddell has continued to spend the winters here, and in the same spirit of devotion maintains the plantation in splendid state and carries on the beautification of the entrance and grounds.

There is about Chicora Wood an air of uniqueness, abetted to some extent perhaps by Mrs. Pringle's writings, that makes it stand out in the hearts of all who visit here. Its fairyland softness and serenity seem to lift one completely out of the realities of the present and into the past of chivalry and gracious living.

Exchange

EXCHANGE PLANTATION, RESTORED AND BEAUTIFIED BY the present owners, is one of the outstanding estates on Peedee. Only a few years ago, the old house was enlarged, the grounds were landscaped, and the whole surroundings transformed into their present loveliness.

The tract of land now known as Exchange can be traced back to the days of the planter Davison McDowell, who owned it as early as 1825.[61] He may have had possession of the plantation as early as 1822, for a McDowell descendant has in her keeping a letter of that year addressed to Davison McDowell, Peedee.[62] One story says that Exchange was so named because McDowell got it in exchange for another piece of property.[63] It is known that he purchased in 1819 from a "Mr. Cohen" the plantation adjoining Chicora. The Rose Bank or Ditchfield tract, as the land also was called, is situated today between Exchange and Chicora Wood.[64] It is possible that this tract figured in the transaction that gave McDowell's plantation its name.

At the age of thirty-nine, Davison McDowell married Mary Moore of Charleston on January 24, 1822. She died the following October on Waccamaw seashore, and five years later her husband married a native of Williamsburg District, Catherine McCrea Witherspoon, widow of Robert S. Witherspoon. The McDowells had six children, among them the Rev. James McDowell, an outstanding and venerated minister of the Presbyterian church.[65]

In 1837, five years before his death, Davison McDowell sold Exchange to Paul Weston, who willed it to Francis Weston. Robert F. W. Allston then bought the 482-acre plantation from Weston in 1843. Allston at that time was beginning the expansion of his holdings. Three years later he sold the property to Cleland K. Huger, who conveyed it in 1853 back to Allston. By that year Robert Allston also had acquired Rose Bank and the nearby Breakwater tract.[66]

Governor Allston, who died in 1864, willed Exchange to his daughter, Elizabeth Waties, the future Mrs. John Julius Pringle, and

EXCHANGE

the future author of *Chronicles of Chicora Wood* and *A Woman Rice Planter*. Elizabeth was to have inherited also some one hundred slaves, but loss of the indebted Allston properties at the end of the War Between the States canceled all provisions of the will.[67]

When the Allston lands were sold in 1869, Robert Allston's two sons, Benjamin and Charles, bought Exchange in partnership. Benjamin, a graduate of the United States Military Academy, had interrupted a rice-planting career to serve in the Confederate army and then had returned to planting during the early post-war years. He finally entered the Episcopal ministry,[68] serving for a time as rector of Prince Frederick's Church. Charles purchased Windsor on Black River in 1885 and made his home there.

In 1899 the Guendalos Rice Company, composed of James Louis LaBruce, Francis Williams Lachicotte, and Louis Claude Lachicotte, obtained part of the plantation from Charles Petigru Allston. LaBruce about 1904 bought all interests of the company and later secured the remaining portion of Exchange. His other plantations acquired through the company were Arundel, Springfield, Enfield, Guendalos, and Birdfield. Two of LaBruce's sons, James Louis, Jr., and George Albert, inherited Exchange on their father's death in 1915. From them the plantation passed to Dirleton's current owner, Thomas G. Samworth, of New Castle County, Delaware, a publisher of books on firearms, who sold it in 1945 to the present owner, Thomas S. Ragsdale, a prominent businessman of Lake City, South Carolina.[69]

To the rear of the one-and-a-half-story house at Exchange, the Ragsdales added a new portion as large as the original. The dwelling ranks with Chicora Wood as one of the oldest houses among Peedee plantations today. The clearing of undergrowth from the grounds by Thomas Ragsdale revealed, besides numbers of live oaks, the outline of an old, formal garden, which Mrs. Ragsdale has restored. To the front, a large artificial lake forms a winter refuge for hundreds of wild ducks.

Now included in Exchange property are the plantations of Rose Bank, once an Allston holding later owned by the Fraser family, and Breakwater, an old Allston place until recent years attached to Hasty Point.[70]

The Ragsdales, who have done so much toward restoring and developing Exchange, live at the plantation all year except for a few weeks each summer.

Hasty Point

HASTY POINT IS SO NAMED, TRADITION SAYS, BECAUSE General Francis Marion made a hurried escape across the Peedee River here from British forces during the American Revolution. With the enemy hot on his trail, so the account goes, the Swamp Fox jumped into a canoe on the shore of the plantation and paddled up a small creek. The British had difficulty finding boats, and when they finally pushed off in the direction they thought Marion had taken, they found themselves in Thoroughfare Creek, which led into Waccamaw River and carried them away from the wily American general.[71]

It is a shame to have to doubt the accuracy of so charming a story; but there is evidence that Hasty Point already had its name some years before Francis Marion eluded his British pursuers. William Coachman, apparently the original grantee, according to one source offered for sale Hasty Point under that name about twelve years prior to the Revolution. He also advertised, along with the plantation, some hundred slaves, stock cattle, hogs, sheep, carts, wagons, and rice tanns.[72]

Ownership of Hasty Point during the early part of the nineteenth century is difficult to trace, since the original papers were destroyed when the house burned a few years ago. Someone of the name Bessard, is known to have held the plantation in 1825;[73] and it is said that Davison McDowell, who had Exchange plantation at one time, purchased Hasty Point about six years before his death in 1842, commenting that he hoped it would not mean the hasty crossing for him that it had meant for Marion.[74] At any rate, Hasty Point did eventually become home for Francis Weston and his family in the years before the War Between the States and remained in the Weston line until 1918.

Francis Weston, born in 1811, married Elizabeth Blyth Hyrne Tucker, daughter of John Hyrne and Elizabeth Allston Tucker of Litchfield plantation on the Waccamaw.[75] The Weston family lived on a lavish scale. Each summer they moved to their lovely home in Charleston, an annual event that led Mrs. Weston to groan once that

HASTY POINT

fifty individuals were taken to and from Charleston every year. Asked how such a large number could be possible, she replied, "'We cannot possibly separate husband and wife for six months; so Harry, the coachman, has to have his wife and children, and the same with the cook, and the butler, and the laundress, until we are actually moving an army every time we move.'"[76]

The war left Francis Weston penniless; the family had Hasty Point but no means with which to plant it. Pauline, the eldest daughter, then about twenty-seven years old, came to the rescue. Taking a small sum given her by an English relative, she invested the money in supplies and opened a small store at Plantersville. By re-investing her income and gradually building up the business, she was able to support her large family and to educate her brothers and sisters. Income from the store also paid off a mortgage on the plantation and made it possible for Pauline to plant rice successfully until a series of freshets made the risk too great. She then returned to her store as the sole means of support for the family at Hasty Point.[77] She died in 1922, but is still remembered for her undaunted spirit.[78]

Hasty Point went out of the Weston family in 1918 with its sale by Pauline to George McCormick, of Atlanta, Georgia, who purchased, too, the adjoining Breakwater plantation.[79] The two places amounted to 1,195 acres. In 1927 George McCormick sold out to Jesse Metcalf of New York City.[80] Expansion of Hasty Point property began soon after Metcalf's purchase and eventually grew to 14,127 acres, the largest holding in the entire section. Among the tracts added on Peedee and Sandy Island were: Belle Rive, once the home of the J. Harleston Read family; Holly Grove, a Tucker plantation later owned by J. Harleston Read; Glenmore, Sandy Knowle, Oak Lawn, Pipe Down, Taylor Hill, Bates Hill, Ruinville, and Oak Hampton. Metcalf owned besides these places the plantations of Willbrook, Oatland, and Turkey Hill, located on the Waccamaw River.

The late H. O. Schoolfield, a well-known and successful businessman of Mullins, South Carolina, bought the Peedee and Sandy Island acreages in 1947 from the Metcalf estate.[81] Afterwards, the new house erected by Metcalf and used by him during his winters burned and was replaced with another dwelling, a long, brick structure on the bluff of the beautiful Peedee River.

Hasty Point, including all the Peedee and Sandy Island tracts purchased from the Metcalf estate, was sold by the Schoolfield family in

April, 1959 to Williams Furniture Corporation of Sumter, South Carolina. Then in March, 1960 Williams sold the plantation house of Hasty Point along with some surrounding acreage, to the present owner, Philip L. Van Every, former mayor of Charlotte, North Carolina, who spends part of each year at this lovely spot on the Peedee River.

Peedee River Addenda

Nightingale Hall

Used for many years since the cessation of rice planting for timbering and hunting, Nightingale Hall fulfills the same purposes today.

After acquiring the property in 1960, Metropolitan Properties, Incorporated, held it until July 14, 1964, when the firm sold out to Williams Furniture Corporation of Sumter, South Carolina. Williams then on December 23, 1964, conveyed a large portion of the combined Nightingale Hall-Clermont-Ingleside tracts to a Mullins, South Carolina, group consisting of E. Smith Buchan, James A. Battle, William H. Daniel, Jr., and William B. Schoolfield. This group ownership continued until 1976 when James A. Battle and the Buchan family sold their interests in the Nightingale Hall property to the group's remaining two, Daniel and Schoolfield. W.H. Daniel, Jr., and W.B. Schoolfield are still the owners of Nightingale Hall.

The house here remains about the same, serving largely as a hunting lodge.

Milton

Milton plantation has remained essentially unchanged over the years. The small house here still serves as a hunting lodge.

Even the ownership of Milton has changed little since the Jericho Corporation got the place in 1961. In 1972, Jericho conveyed the tract, at this time 350 acres, to three individuals, Wallace D. Gibbs, Jr., R. Stuart Dickson and J. Lewis Bibb, all of Charlotte, North Carolina. Wallace D. Gibbs died in 1987. His interest is at this time held by the Gibbs family. R.S. Dickson and J.L. Bibb remain the other owners.

Dirleton

This rice plantation is now a wildlife preserve owned by the state of South Carolina. Consisting at the time of 353 acres of riceland and 271 acres of upland, Dirleton was deeded on August 4, 1961, by

Thomas G. Samworth to the South Carolina Wildlife Resources Department, now known as the South Carolina Wildlife and Marine Resources Department.

The state's purpose in accepting Dirleton plantation was to use it "as a public waterfowl management area," the supervising agency claims. This work continues on a year-round basis today. The interesting, old Dirleton house, a superb example of typical Lowcountry plantation architecture, contains offices for the plantation's operations.

Mr. and Mrs. Thomas G. Samworth retained life tenure of the property and continued to live here until their deaths. She died in 1974, and he died in 1981. Both are buried amidst tall magnolias on the grounds south of the house. The Samworths deserve recognition for their generous gift of Dirleton to South Carolina for the advancement of wildlife conservation.

Arundel

The most notable development at Arundel during the last thirty years or so has been its beautification by Mr. and Mrs. Alexander McQueen Quattlebaum, who became owners of the plantation in 1955. The Quattlebaums added the columns to the river entrance of the house and the handsome paneling to portions of the interior. To the naturally beautiful grounds already blessed with large live oak trees and already planted with banks of azaleas, they added more shrubbery. The construction of brick walks all through the grounds completed the landscaping. Under the Quattlebaums' loving attention, Arundel has not only flourished, it has become a showplace among Georgetown plantations.

Alexander Quattlebaum died on March 10, 1987, at Arundel, which remains today the home of his widow, Lucille Godfrey Quattlebaum.

A prominent civil engineer, business executive, philanthropist and supporter of education and the arts, Alexander Quattlebaum was a distinguished South Carolinian whose accomplishments were many. To enumerate just some, he earned bachelor of science, master's and doctoral degrees in civil engineering and taught civil engineering for a while at Clemson College. He went on to organize Harlee-

Quattlebaum Construction Company of Florence, South Carolina, serving as president and later chairman of the board. Among his various board memberships were those on the boards of trustees of Clemson University (formerly Clemson College), Presbyterian College, Brookgreen Gardens, Spoleto Festival U.S.A. and the South Carolina Historical Society. He also served on numerous professional, business and civic boards. Moreover, he made generous contributions to the causes he served. He endowed engineering professorships and a faculty achievement award at Clemson, honor scholarships at Presbyterian, established the Gian Carlo Menotti Artists-in-Residence Endowment at the College of Charleston and the Quattlebaum apprenticeship program for College of Charleston students in the Spoleto Festival.

Since her husband's death, Lou Quattlebaum, an artist and book illustrator, has continued support of cultural activities in the state.

Through the years, the Quattlebaums have enhanced not only the beauty of Arundel but the quality of life, through their altruism, for many citizens of South Carolina.

Prince Frederick's Church, Peedee

The complete collapse of Prince Frederick's Church, Peedee, as a congregation and the dismantling of its pre-Civil War structure have been the major developments here in the last quarter-century. Very little of a positive nature can be said about the situation. Once services were transferred from here to the summer chapel in Plantersville, maintenance on this building ceased. Finally, in the mid-1960's, the Episcopal Diocese of South Carolina tore it down and sold off the material. Only the tower and front wall were left to mark the site. Some years later, the congregation ceased to exist.

One cannot help but wonder about the inevitability of such a fate for Prince Frederick's Church. Little effort and interest seemed to be shown during the 1960's by the diocese and the community in preserving at least the lovely old church building if not the organization. No leader seemed to emerge from anywhere to promote the cause. Today, the church's tower and front wall serve as a monument to apparent indifference.

Chicora Wood

Renowned Chicora Wood today serves as a wonderful example of a local plantation that is not only intact but meticulously restored to its pre-Civil War handsomeness. It is the remarkable accomplishment of the present owners, Mr. and Mrs. Jamie Wyndham Constance of Santa Barbara, California. No expense and no effort have been spared by the Constances in their conscientious endeavor to preserve and to restore not only the fine, old house but the various outbuildings as well. In addition, they have replanted the grounds. The results of such keen appreciation are obvious. Chicora Wood surely has never been in better condition or looked more beautiful than now.

The long Waddell ownership began in 1926 when Duncan Cameron Waddell, Jr., bought the property from Elizabeth Allston Pringle's estate. Following her husband's death, Mrs. Waddell continued to live at Chicora Wood. Upon her death in 1961, her nephew T.S. Andrews inherited the plantation. The Andrews family lived here for several years, then moved to Georgetown. Finally, on March 15, 1984, T.S. Andrews sold the property to Mr. and Mrs. Constance.

In addition to restoring Chicora Wood's house, the Constances have conducted diligent research into its history and have, as a result, brought to light some significant findings. For instance, the back wing was once the south wing to the first plantation dwelling constructed during the 1740's under the ownership of John Allston, one of the two Allston brothers to establish the family on Waccamaw Neck. That first house shown on an old plat consisted of a center section and two wings, all at the edge of the Peedee River and facing the water. By the time Mrs. Benjamin Allston moved from Waverly to Chicora Wood in 1819, something apparently had happened to the north wing and the center portion. The first phase of the present house, it is speculated, was placed about this time in front of the remaining south wing and attached to it by passageways. The second phase was an addition to the front part in 1838. This included the piazza and main entrance as they are today.

Of outstanding historical importance, Chicora Wood offers an unusually well-preserved, vivid illustration of a Carolina rice plantation. Moreover, it is once again a working plantation. Turf is the product now. Large amounts of turf are grown and sold commercially here each year. Too, Chicora Wood has retained its old beauty and charm, truly a very special place.

The plantation was added in 1973 to the National Register of Historic Places.

Exchange

Probably the most noteworthy thing about Exchange today is that it has remained relatively unchanged. In this instance, lack of change means that the plantation is still intact and still a lovely place.

Buying the property in 1945, Mr. and Mrs. Thomas S. Ragsdale of Lake City, South Carolina, resided here the rest of their lives. Mrs. Ragsdale died in 1974 and Mr. Ragsdale in 1977. The plantation then passed eventually to Mr. and Mrs. Thomas S. Ragsdale, Jr., also of Lake City. Thomas, Jr., died in 1991. Since then Exchange has been owned by his wife and children.

Nicely maintained through the years, Exchange owes its preservation to the faithful attention of two generations of the Ragsdale family.

Hasty Point

Hasty Point is another plantation that has remained unchanged in recent years. It is still intact, and the grounds are attractive. An impressive brick gateway is at the head of an extensive live oak avenue planted in this century. Near the house and along the river bank is a grove of older, larger oaks.

Philip L. Van Every of Charlotte, North Carolina, bought the plantation, containing at the time 500 acres of upland and 200 acres of rice fields, in 1960 from Williams Furniture Corporation of Sumter, South Carolina. Mr. Van Every then sold the same acreage on February 15, 1979, to Robert P. Schofield, Jr., who made Hasty Point his year-round home. On July 22, 1980, Mr. Schofield sold the plantation to his son Robert P. Schofield, III, but he and Mrs. Schofield have continued to reside in the Hasty Point house.

Not only can this plantation, one of the oldest in the area, boast of the many live oaks here, it can still boast—as always—of its scenic, sweeping view of the river and of Thoroughfare Creek, a major waterway to the Waccamaw River. Without a doubt, Hasty Point remains one of the choicest settings on the Peedee.

SAMPIT RIVER AND WINYAH BAY

Friendfield

TODAY ONE OF THE MOST IMPRESSIVE AND BEAUTIFULLY landscaped plantations in the low country, Friendfield was for many, many years the possession of the Withers family. As early as 1769, this Sampit River plantation was the home of a Withers. James Withers and his wife, the former Mary Esther Stewart, were living in the original dwelling, a small brick house, when their son Francis, a later owner, was born on July 14 of that year.

Friendfield was originally named Washington—conceivably for the President. When Francis Withers bought the plantation, he lacked sufficient funds to complete the transaction and so borrowed from one of his brothers. When the money was returned, the brother refused payment and tore up the note. This act of generosity prompted Francis to rename the plantation Friendfield.

Francis Withers attended Harvard from 1788 to 1792 and later married Elizabeth Thomas, daughter of Edward Thomas. A year after Elizabeth's death, he married Mrs. Sara Warham, widow of William Warham and a first cousin of his deceased wife. Coming to Friendfield with the second Mrs. Withers were her small daughter, Elizabeth Warham, her mother, her first mother-in-law, and Mrs. Withers' sister.

In 1818 Francis built the large mansion so noted over the whole countryside. The front steps and porch were of flagstone; the banisters and stair railing were made of wrought iron. Wallpaper, described as a "work of art," is said to have been brought over from Paris. An outstanding feature put in by Francis Withers was a circular stairway extending from the basement to the attic. Legend claims that the old house had a secret passage leading to the Sampit River, but the last occupants of the house before its burning never found such a passageway.[1]

Before his death in Charleston in 1847, Francis Withers had acquired a number of plantations, including Northampton, Mount Pleasant, Canaan, Midway, Westfield, Harmony, and Bonnyneck, all on the Sampit,[2] and Springfield on the Waccamaw.

FRIENDFIELD

A close kinsman, Robert Withers, who married Esther Johnstone, lost his whole family in a hurricane that swept in unexpectedly during the night of September 27, 1822, and drowned about forty summer residents on North Island. Robert Withers and three Negroes were the only survivors of eighteen people occupying his house. Mrs. Withers, her four daughters, and her son perished. Robert Withers' account of the storm tells of the household's removal between one and two o'clock in the morning, after the chimneys had fallen, to a safer room on the west side of the house. Within ten minutes the building wavered, and an instant later the whole structure "crumbled like an egg shell." He remembered falling down and finding himself in the water and hearing a call to catch one of his daughters. Then his consciousness failed. Hours later, when he came to himself he was hanging to a piece of timber with his servant on the other end. He heard a voice behind him and turned to see his twelve- or thirteen-year-old son clinging to wreckage. Only a few minutes afterwards, a wave dashed masses of debris on them all. The little boy was never seen again. In a short while Robert and the Negro were pulled from the water by a young naval lieutenant who had seen them from the shore. A Negro boy from the Withers house, who was washed across to Debordieu Beach on a pair of steps, and an older man, who suffered a broken arm when the house collapsed, both survived the disaster. Robert Withers never knew how his family had been killed—whether crushed to death in the house or washed out to sea.[3]

Elizabeth Warham, Francis Withers' stepdaughter, eventually married Dr. Alexius M. Forster, and the couple made their home at Friendfield. When Francis Withers later fell into ill health, he turned over the plantation's management to his son-in-law, and upon his death the property was left to Mrs. Forster. During the War Between the States, Dr. Forster served as an army surgeon, practicing afterwards in and around Georgetown. He died suddenly from injuries received when his horse bolted one day and threw him from the buggy.[4]

In 1897 Dr. Forster's widow and the rest of his family sold Friendfield, consisting then of 1,200 acres, to B. Walker Cannon, of Georgetown,[5] from whom the property passed in 1919 to Hugh L. Oliver, of Georgetown, who conveyed it the same year to Patrick C. McClary, of Williamsburg County.[6] It was during McClary ownership that the old Friendfield house was renovated—only to go up in flames in 1926, just a few years before the present owner purchased the plantation.

Several of the quaint slave cabins and the old chapel burned about 1936. Another landmark, the rice mill, was torn down between 1935 and 1940.[7] After Friendfield's ownership by a hunting club of Florence, South Carolina, the plantation was acquired in 1930 by the current owner, Radcliffe Cheston, Jr., of Philadelphia, Pennsylvania.[8]

Two years later Cheston erected the present Friendfield house, a handsome, imposing mansion, now the seat of a vast estate which

COLONNADE OF FRIENDFIELD HOUSE

includes about 3,500 acres, encompassing besides Friendfield, the plantations of Canaan, Bonnyneck, Silver Hill, and Midway. In developing his property, he altered the arrangement of the Friendfield grounds, changing the driveway from the front of the house to the rear and laying out the lovely camellia garden. Magnolia trees were planted on each side of the entrance for an impressive distance; later, when they reach great heights, they will further enhance the plantation's stature as a showplace. The beautiful sunken garden behind the residence was laid out and planted long ago by slave labor.[9] South Carolina's first tea plants are supposed to have been planted here; and among the varied shrubbery were Norwegian pine, mountain laurel, and varnish trees.[10]

Maryville

MARYVILLE WAS HELD FOR MORE THAN A CENTURY BY the Read family before passing into the hands of other owners in 1939. The greater portion of the plantation now is absorbed in an incorporated town of the same name, but the old house and grounds which form the year-round home of the present owner, Mrs. Charlotte Ives Montgomery, are untouched as yet by urbanization.

Maryville plantation is situated between the mouth of Sampit River and Winyah Bay. Records trace its ownership during the closing years of the eighteenth century to the Withers family, when the property was owned apparently by James Withers, a brother of the Francis Withers who owned Friendfield farther up the Sampit. Large acreages on both sides of the Sampit River belonged to the Withers connection, and it was from a south bank tract that Maryville was established. This particular piece, from all indications, was the inheritance of James Withers' niece Mary, for whom the plantation is named; and it was through her marriage in 1811 to John Harleston Read (born in 1788), son of William and Sarah Harleston Read of Charleston, the land came into the Read family. It was to remain so for the next 128 years. Shortly after coming into possession of Maryville, John Harleston Read purchased neighboring Oakley, Lucerne, and Upton plantations, combining these with Maryville to make up one tract of 1,800 acres.

In 1811 Read established his home on the Maryville tract and built the residence that stands today. This substantially constructed dwelling went unchanged from the time of its erection until 1890, when John Harleston Read III, added a large kitchen and storeroom to the west side. After Mrs. Montgomery became owner, this wing was sold; it now serves as a complete house for a family in the town of Maryville.

Mary Withers Read died May 9, 1817, and on April 17, 1822, John Harleston Read took a second wife, Emily Ann Huger, daughter of Major Benjamin Huger. In addition to running his plantations, Read

MARYVILLE

(House no longer exists)

for many years represented Georgetown District in the General Assembly of South Carolina. He died in 1859, leaving Maryville to his eldest son, John Harleston Read II, born in 1815.

The second John Harleston, who married Esther Jane Lance in 1840, had already established his home at Belle Rive plantation on Peedee at the time of his father's death. He came to own besides Belle Rive and Maryville, a summer home at Plantersville, and a house in Charleston. In addition, he and his brother, James Withers Read, inherited Harmony plantation on the Sampit from his greatuncle, Francis Withers.

Originally called Smurden, Belle Rive received its present name in 1855, following a trip by John Harleston Read II to Belle Rive school for boys at Vevey, Switzerland, where his young son, John Harleston III, was entered as a student. Read was so taken with the ring of the name that he decided to give it to his Peedee property.

Though opposed to secession, John Harleston II volunteered for military service at the beginning of the War Between the States and recruited and equipped at his own expense a company composed of Georgetown District men. This company, with Read as captain, became Company A, 21st Regiment, South Carolina Volunteers, Hagood's Brigade, Army of Northern Virginia. Captain Read, later promoted to major, commanded the Georgetown company in many of the Virginia battles. He returned to Belle Rive at the end of the war and died in 1866 in Charleston.

Upon his father's death, John Harleston Read III assumed ownership of Maryville and planted rice there and at Belle Rive for several years. He also managed plantations for his mother and for his grandfather, the Rev. Maurice Lance. He later conveyed Maryville to his uncle, Benjamin Huger Read, who lived at the plantation until he died in 1887. In 1889 John Harleston Read III re-acquired Maryville from Benjamin H. Read's estate and moved his family there to live. After his marriage in 1870 to Ann Elizabeth Stoney, Read and his wife had resided at Belle Rive and at Plantersville. They moved to Maryville and were still living there when John Harleston Read III died in 1912. Read had been active in civic affairs about Georgetown during his lifetime; particularly noteworthy was his service as a delegate to the South Carolina Constitutional Convention of 1895 and as first commander of the Arthur Manigault Camp of Confederate Veterans, which he organized.

Following the death of John Harleston Read III, his daughter, Elizabeth Read Wood, the wife of Alan Wood III, of Philadelphia, Pennsylvania, bought out the interests of the other heirs to become owner of Maryville.[11] The Woods lived at the plantation for only a few years before selling it on January 24, 1939, to a real-estate company.[12] Thus had the long reign of the Read family at Maryville come to an end. John Harleston Read III, before he died had the pleasant experience of being visited here by his grandson, John Harleston Read V—just as in his own childhood, he had stayed at Maryville with his grandfather, John Harleston Read I.[13]

In March, 1939, the late Colonel Reading Wilkinson, one-time district army engineer at Charleston, purchased the Maryville house and a small surrounding acreage.[14] The greater part of Maryyville land was going at this time into development of the town of Maryville. Colonel and Mrs. Reading Wilkinson made their home at the plantation until its sale by them in 1944 to the present owner, Mrs. Charlotte Ives Montgomery, daughter of Colonel and Mrs. Robert L. Montgomery of Mansfield plantation on Black River.[15] Colonel and Mrs. Wilkinson later purchased Hopsewee plantation on North Santee, Mrs. Wilkinson's current home.

Belle Isle

IN THE EARLY YEARS OF THE NINETEENTH CENTURY Belle Isle was the plantation of a celebrated Revolutionary officer; during the War Between the States it served as a harbor battery; and today its grounds form one of South Carolina's outstanding gardens.

The plantation was once part of the vast Winyah Barony, a tract of 12,000 acres granted to Landgrave Robert Daniell in 1711. After one day's ownership of the Barony and other lands that amounted altogether to 24,000 acres, Daniell sold to Landgrave Thomas Smith, second landgrave of the name. Landgrave Smith, on his death in 1738, devised 1,000 acres of Winyah Barony to each of his four surviving sons, Henry, Thomas, George, and Benjamin Smith. Since George died under age and unmarried, his 1,000 acres were divided among the other brothers, leaving each with an approximate 1,333 acres. On March 25, 1756, Henry Smith sold his 1,333 acres to Elias Horry II; and a year later Horry purchased 346 acres of Benjamin Smith's share.[16] It was from most, or all, of these two tracts that Elias' nephew, General Peter Horry, formed his three Winyah Bay plantations, Prospect Hill, Dover, and Belle Isle.[17] In 1801 Peter Horry laid out and designated 640 acres of his land as Belle Isle, and in that year the property still formed "a part of Gen'l Peter Horry's plantation named 'Dover.'"[18]

The Horry family was established in South Carolina by Elias Horry, a French Huguenot, who came to America about 1691. He married Margaret Huger, daughter of Daniel Huger, a settler on South Santee River,[19] and had by this union, six children, Daniel, Elias II, Peter, John, Margaret, and Magdeline.[20] John's son Peter was to become one of the state's prominent figures during the Revolution.

In 1775 the Provincial Congress of South Carolina elected Peter Horry with nineteen other men as captains to serve in the First and Second South Carolina Regiments. Assigned to the Second Regiment, Horry was promoted the following year to major. In 1779 he received a promotion to the rank of lieutenant-colonel and was attached to the

Fifth Regiment. The next year, the First, Second, Third, Fifth and Sixth regiments were consolidated into three. Horry at that time was placed on a list to await a vacancy in his rank. Later in the year he was assigned to duty with the South Carolina militia, as Governor Rutledge made Francis Marion (whose regiment had been taken at the fall of Charleston while he was on furlough) brigadier-general of the lower brigade of South Carolina militia. Horry became a colonel of regiment under Marion and afterwards organized a regiment of light dragoons for state service. In this capacity he stayed until his regiment was combined with another near the end of the war and put on Continental Establishment.

Peter Horry later wrote a history of Marion's brigade, turning the work over to Mason L. Weems for structural improvement; but it is claimed that the facts were distorted by Weems, who acknowledged having written a military romance in his *Life of Marion.*

Not long after the Revolution, Horry was appointed brigadier of the South Carolina Militia's Sixth Brigade, so serving until 1806. In 1801 a new district was carved from Georgetown lands and named Horry in honor of the General, whose brigade included the militia of that area. This district is the Horry County of today—more characteristically dubbed "The Independent Republic of Horry."

General Horry was married in 1793 to Margaret Guignard, daughter of Gabriel Guignard, a French Protestant. He died in Columbia on February 28, 1815, at the approximate age of sixty-eight and is buried in the churchyard at Trinity Protestant Episcopal Church.[21]

By provisions of Horry's will, Belle Isle went to his widow. Dover was devised to his niece, Mary Horry; Prospect Hill was left to Sarah Hall Horry Bay, and a fourth place, Cover plantation on Santee, was given to Peter Horry Joor, son of John Joor.[22] Mrs. Peter Horry, at her death, willed Belle Isle to Sarah Hall Horry Bay,[23] to whom Prospect Hill already belonged. Sarah later married William Mayrant, Jr., son of Colonel William Mayrant, a planter of Sumter District. Her eldest child, Sarah Ann Mayrant, born in 1820, was married in 1844 to William Eveleigh Richardson,[24] and the Richardsons' first son, William Mayrant Richardson, was given Belle Isle plantation by his grandmother, Mrs. Mayrant, in 1868.[25] William Richardson died in 1888,[26] and the estate came into the possession of Gardner B. Penniman of Georgetown in 1894.[27] Mrs. Penniman is given the credit for having begun the famous gardens here about 1895.[28]

BELLE ISLE

During the War Between the States, the plantation served as the site of Battery White, which guarded the entrance to Georgetown Harbor. Near here on the morning of March 1, 1865, "The Harvest Moon," flagship of Rear Admiral J. A. Dalhgren, of the United States Navy, was blown up by a torpedo placed in the channel of Winyah Bay. The vessel is said to have sunk within five minutes. Later the material rescued from the wreckage was taken to Battery White. Captain Thomas W. Daggett, commander of coastal defenses from Little River to Georgetown, was responsible for this major accomplishment. The deep trenches overlooking the bay still are here, and resting within the breastworks are cannons left as they were overthrown by Federal forces at the end of the war. This interesting terrain is included now in the gardens.[29]

SITE OF BATTERY WHITE, BELLE ISLE

Gardner B. Penniman in 1919 sold Belle Isle, containing its original acreage, to one of the county's most prominent and wealthy rice planters, B. Walker Cannon, as trustee. Transferred, too, were 250 acres of Mount Hope and the tracts of Dover and Prospect Hill.[30] Mrs. Cannon conveyed the Winyah Bay portion of Belle Isle in 1928 to Frank E. Johnstone, descendant of an old Georgetown rice-planting family.[31] Under Johnstone's ownership, Belle Isle Gardens was opened to the public.

In 1929 Mrs. Henry M. Sage of Albany, New York, who was seeking a winter home and was attracted by the mild climate at the bay, secured a ten-year-lease on the plantation. The estimated cost of altering the existing Belle Isle house was so great, it was found, that moving another house here would be more feasible. A friend suggested the Mendenhall place at Newberry, South Carolina. Considered one of the state's fine old homes, this house had been constructed about 1829 for Dr. Marmaduke Thomas Mendenhall, who lived in Newberry until 1837. The house, which contained a third-floor garret in addition to the main and second floors, was taken down and brought to Belle Isle and then re-erected amid the colorful gardens on Winyah Bay. A few changes were made to bring it up to modern standards of comfort: the structure was raised to provide space for a kitchen, servants' rooms, and a furnace room on the ground floor; the wing originally at the back was placed at one end and balanced by a similar one at the other end; and closets were installed.

One of the most outstanding features of the old house was the 25-by-40-foot parlor to the left of the long hall. Extending the length of the dwelling, the room had scenic wallpaper depicting, on one side, a skirmish between two bands of Indians and, on another part, a group of young colonial couples against a background of lake, trees, and flowers. The figures of this delightfully quaint paper, so large as to be almost life-size, reached from the four-foot wainscoting to the ceiling.

A gracefully designed stairway reached from the main floor to the garret, and ten steps above the main level there was a landing across the hall. On the outside of each step were scroll-relief carvings; while on the base of the landing there were carvings of "palm-like designs." Providing a final decorative touch was the railing made entirely of mahogany.

Mrs. Sage's efforts to buy and move the house at the expiration of the lease failed, but she had reserved the right to remove the wallpaper and staircase—both of which now are in her splendidly restored residence at Dover plantation farther down Winyah Bay.[32] The lovely Belle Isle house burned to the ground just a few years after the Frank Johnstones made it their home, and today only brick ruins of the chimneys and foundations remain to tell of the mansion's former grandeur.

Since Frank Johnstone's recent death, Belle Isle Gardens has been under the management of his son, William A. Johnstone, who opens the grounds to the public each season. Belle Isle, aside from its historic significance, is famous for its azaleas and camellias and its superb panoramic view of broad Winyah Bay.

Dover

DOVER, ONCE PART OF ANCIENT WINYAH BARONY, IS TO-day one of the most impressive estates of the South Carolina low country. Its development in recent years has brought into being, perhaps more so than did the rice-planting era, this plantation's "golden age."

The owner, Mrs. Henry M. Sage, of Albany, New York, who re-erected at Belle Isle the Mendenhall house from Newberry, placed at Dover in 1949 the enormous dwelling from Woodlawn plantation in Berkeley County.[33] Described at its original site as the finest and largest residence between the upper Cooper and Santee rivers,[34] this house was re-erected, with some alterations to provide for modern living, in the new setting on Winyah Bay. Woodlawn plantation rests now beneath the waters of Santee-Cooper lake; the house was saved only by the miraculously good fortune of its removal to Winyah Bay.

Acreage included in Mrs. Sage's estate comes actually from two tracts, the Dover plantation of ante bellum days and adjacent Oakton plantation, a holding of subsequent development that appears strongly to have been cut off from an originally larger Dover. A large part of ante bellum Dover is attached now to Rice Hope plantation, but Mrs. Sage prefers to call her place Dover, rather than the more recent name Oakton, and the north boundary of her property does lie within the Dover plantation of early years.

Named for the adjoining Dover landing, Dover plantation stemmed originally from Winyah Barony,[35] a tract of 12,000 acres granted on June 18, 1711, to Robert Daniell. After one day's ownership, Daniell conveyed the barony and other property, totaling in all 24,000 acres, to Landgrave Thomas Smith, second landgrave of the name. Because of its long ownership by the Smith family, the barony often was called Smith Barony. Thomas Smith, the second landgrave, came to Carolina in 1684 with his father, but there are no records to show what he did with his Winyah Bay property before 1732. In that year, at least, he advertised for sale 14,000 acres of the 24,000 acquired

DOVER

from Robert Daniell. He sold none. Five years later, encouraged apparently by the growth of Georgetown, he tried to establish in competition a new town on his land. This time he advertised the lots in his proposed town, making special inducements to attract settlers and offering lots to ministers who might desire to live there. However, the Landgrave was able to sell only few lots, if any; for, a year and a half later, he placed still another advertisement in the *South-Carolina Gazette,* adding further inducements. Despite these elaborate plans, no map of Smith's proposed town has been found, and there is no record of purchases of lots by those who urged the project.[36]

Thomas Smith eventually gave 3,000 acres of Winyah Barony to his eldest son, Thomas, whose will of December 3, 1729, devised to his sister, Justinah Moore, 1,000 acres from the 3,000. Justinah, after her brother's death, advertised her inheritance for sale. Surviving his eldest son, Landgrave Smith married again and by his second wife had another son whom he named Thomas. His other children were George, Henry, Benjamin, Anne, Justinah, Sarah, and Elizabeth. Smith died on May 9, 1738, and left to each of these children two lots in the town he was still trying to settle and 2,000 acres of other land "'to make each of them a small Retiring Country Seat.'" The remainder of the Barony, extending north and south from the area of the proposed town, was bequeathed to his sons, Henry, Thomas, George, and Benjamin, giving each of these another 1,000 acres apiece. After the Landgrave's death, the fruitless efforts to settle Smith-Town ceased, and Georgetown continued as the port town of the upper coastal region.[37]

The son, George Smith, died under age and unmarried; so his 1,000 acres were divided among the other brothers, making them each have an approximate total of 1,333 acres. Henry sold his share on March 25, 1756, to Elias Horry, and Benjamin, a year later, sold 346 acres of his part to the same purchaser. The rest of Benjamin's portion remained in his family for many years, passing from him to his eldest son, Thomas, who in turn left it to his eldest son, Thomas John, after whose death in 1834, the remaining part held by him and forming Retreat plantation was sold away. Barony lands held by other members of the Smith family eventually were sold to other owners.[38]

Elias Horry's land acquired from Henry and Benjamin Smith came into the possession of Elias' nephew, General Peter Horry, who formed from it the plantations of Prospect Hill, Dover, and Belle Isle,[39] the last being created in 1801 by the laying-off of 640 acres from Dover property.[40] Prospect Hill and the greater part of Dover seem to have been coupled in conveyances from the late nineteenth century through their sale in 1935 to William N. Beach of Rice Hope plantation.

General Horry willed Prospect Hill to Sarah Hall Horry Bay, and Dover to his niece, Mary Horry. Belle Isle was left to Mrs. Horry,[41] who willed it at her death to Sarah Bay.[42] Sarah married William Mayrant, Jr., son of Colonel William Mayrant, of Sumter District;[43] and Dover plantation finally passed to her following its ownership by Mary Horry.[44] The Mayrant family thus obtained control over General Horry's three places, Prospect Hill, Dover, and Belle Isle.

Oakton plantation, adjoining Dover on the south, is known to have belonged to William Mayrant at one time.[45] Indications are that Oakton was part of Dover when inherited by the Mayrants from General Horry and was cut off to form a separate plantation. William Mayrant died in 1840.[46] Oakton, containing 800 acres, became the lands of Sampson Allston, a Negro, in 1899, and after he died, it was divided among his children.[47] However, some one hundred acres bordering on the bay came into Frank E. Johnstone's possession and was conveyed by him in 1913 to Wayne Darlington,[48] who constructed a house here that was torn down after Mrs. Sage's purchase.[49] Darlington included this 100-acre tract in his sale of Annandale and Hopeland plantations in 1919 to Charles M. Reeves of Sanford, North Carolina.[50] It was purchased by Mrs. Sage in 1931.[51]

On March 23, 1844, William and Sarah Bay Mayrant's eldest child, Sarah Ann, was married at Dover to William Eveleigh Richardson.[52] Thomas Eveleigh, second son of William and Sarah Ann Richardson, ultimately came to own Dover and Prospect Hill, selling both in 1887 to Samuel Sydney Fraser of Georgetown as trustee. At this conveyance the two tracts contained 1,549 acres.[53] In 1906 Fraser's heirs sold the property to Frank E. Johnstone,[54] from whom it transferred a year later to Gardner B. Penniman, owner of Belle Isle.[55] Penniman then sold Prospect Hill, Dover, Belle Isle, and part of Mount Hope to B. Walker Cannon in 1919.[56] With the exception of Belle Isle Gardens and a portion of Mount Hope, the property was acquired from Mrs. Cannon by the late William N. Beach in 1935.[57]

Erection of the present house at Dover began in 1949. It consists, actually, of parts from three different houses: the residence from Woodlawn plantation, with certain alterations; the main doorway, Palladian window, and balcony from the Hunter house of Savannah,

Georgia; and the wallpaper and staircase from the Mendenhall house of Newberry.[58]

Restoration of Woodlawn plantation house at its new location on Winyah Bay is the fulfillment of a desire long held by Mrs. Sage. When she first saw the mansion at Woodlawn in 1914 during a shooting trip with her husband, the late Henry M. Sage, the house had been so reduced from the glories of former days that it then was being used for the storage of cotton bales, cow peas, and similar produce. She became interested immediately in the old dwelling, but following the advice of others who doubted its practicability as a winter home, she did not buy it. During this same visit Mrs. Sage tried to purchase a fine Hepplewhite sideboard at the plantation, but the owner refused to sell, giving as his reason his need of the piece for storage of cotton seed.

Mrs. Sage took pictures of the decayed mansion, and in the 1930's when the Santee-Cooper hydroelectric project was getting underway, she returned to Berkeley County in search of the place she had seen twenty years earlier. The condition of the house had not improved, nor had a more elevating use than storage space been developed for its lovely interior. The owner, after much deliberation, decided to sell; and in August, 1940, Mrs. Sage had the residence removed to her Dover plantation on Winyah Bay, where the various pieces were stored until the house could be re-erected.[59]

Containing numerous closets and store rooms in addition to seventeen large rooms, the Woodlawn house was constructed during the first decade of the nineteenth century by Stephen G. Deveau, a close relative of General Francis Marion. A newspaper article of 1935 describes the structure as "perhaps, the most imposing of all the historic old houses in St. John's parish to be demolished to make way before the diverted tide of Santee, which threatens to sweep over much of this section with the development of the Santee-Cooper hydro-electric project." After the death of Stephen G. Deveaux, his son, Stephen L. Deveaux, sold Woodlawn to Dr. Henry Ravenel, one of the largest landowners in the parish. There, in the spacious plantation house, the Ravenel family lived for a number of years. Dr. Ravenel later gave the plantation to one of his sons, William F. Ravenel, whose heirs sold it to Percival Ravenel Porcher, from whom Clarence P. Gourdin of Pineville, a later owner, obtained possession.[60]

The laborious task of re-assembling and re-erecting this enormous house at Dover began in 1949. Portions of the structure that had deteriorated—such as the ballustrade above the roof and pillars of the piazza roof that once extended far over the porch proper—were replaced. The woodcarving, patterns of which vary from room to room, was added to the residence as were the wallpaper and staircase saved from the Mendenhall house at Belle Isle.[61]

Handsomely restored in every detail, the house, as now arranged and furnished, is unsurpassed by any old plantation home in the state. Despite immense spaciousness it conveys an atmosphere of warmth and softness not always achieved in large dwellings. Many plantation mansions of today, though imposingly styled, have a cool aloofness successfully avoided here.

Mrs. Henry M. Sage's purchase and restoration of the Woodlawn house have been a monumental contribution to the preservation of South Carolina history. Waters of Santee-Cooper dam have submerged the lands of Woodlawn, but the plantation lives on in its proud mansion that stands on Winyah Bay.

Estherville

ONE OF THE MOST SUPERBLY LANDSCAPED ESTATES IN the entire county, Estherville is also one of the oldest. It was formed as early as the mid-eighteenth century from lands of the old Winyah Barony.

The plantation is owned now by George L. Buist, prominent Charleston attorney, who maintains the house and grounds in excellent condition. A lovely natural setting here has been used to full advantage in developing the gardens and lake. At springtime the eye-filling masses of azaleas contrast with ancient live oaks trailing Spanish moss from their huge limbs and are reflected below in the waters of the pond once used to flood the rice fields.

Estherville, as George Buist purchased it in 1943, included 3,000 acres.[62] Its history goes back to the first days of settlement on Winyah Bay and to Winyah Barony, at first the possession of Robert Daniell and then of Landgrave Thomas Smith.[63] The Estherville portion of Winyah Barony came into Johnstone ownership sometime during the mid-1700's when Andrew, son of Archibald Johnston (who spelled his last name without the *e*), purchased it from Robert Smith. Andrew's mother was formerly Esther Allston, and it was for her that Estherville is said to be named,[64] though Esther Johnstone Withers, who lived at a later time, may have been the one so honored.

Herbert Ravenel Sass writes that in 1758 Estherville was the plantation of a McKewn Johnstone, who, by that time, "had demonstrated the practicability of using the fresh water tides in the Low Country's rivers to irrigate broad new rice fields made by dyking off and clearing the vast river swamps."[65] Another authority gives the location of this experiment as "near Winyah Bay,"[66] leaving open the possibility that the site was North Santee River rather than the actual Winyah Bay area.

Former rice planters of Georgetown County have expressed doubt that such an experiment with fresh-water tides could be carried out

ESTHERVILLE

successfully at Estherville in view of its proximity to the ocean. As they recall the days of planting rice, fields here were flooded from reserves. Tide-water so near the ocean, they maintain, was too salty for use in irrigating rice fields. The likelihood is greater that a successful experiment of this kind would have been conducted at a point farther up the bay or on North Santee. Whatever the circumstances and location of this first experiment, the tidal system was a tremendous advance over the reserve method of flooding rice, and within a half-century after its initial use, this new method had to a great extent supplanted the older reserve operation.

In tracing the ownership of Estherville, it is difficult to place a McKewn Johnstone here in 1758, if one follows the data given by the Johnstone family. Their information states that Archibald and Esther Johnston had one son, Andrew, to live to maturity. Andrew was married in 1773 to a Sarah MacKewn and lived at Estherville.[67] He owned also another plantation, Millbrook, on the North Santee, known today as Annandale, where the first tide mill for processing rice was constructed for him in 1791 and 1792 by the celebrated inventor Jonathan Lucas.[68]

A son, William, and a daughter, Esther, were born to Andrew and Sarah MacKewn Johnstone. William planted rice at Millbrook and operated the rice mill there. It is believed that his sister, Esther, and her husband, Robert Francis Withers, may have owned Estherville at one time, even though William originally inherited the place. Indications are that Robert and Esther Johnstone Withers did not live at Estherville but at Millbrook with William Johnstone from about 1812 until 1822. The Estherville house is thought to have been burned by this time.[69] In 1822 Esther Withers and all of her children perished in the severe hurricane that demolished North Island.[70]

Francis Withers Johnstone, a son of William, eventually came into possession of Estherville plantation and sold it in 1853 to his nephew, William, a son of the Andrew Johnstone, who was living then at Annandale,[71] formerly Millbrook plantation. For a time during and after the War Between the States, Estherville passed out of the family's ownership, and William Johnstone planted rice on Black River.[72] However, he regained his Winyah Bay estate in the name of his wife, Alice, in 1878,[73] and planted here until his death in 1893. In 1902 Mrs. Johnstone sold out to her son, Frank E. Johnstone, who planted rice at Estherville for some years, and who sold the place finally in 1918 to John A. Miller of Pennsylvania.[74]

The present house, built originally as a bungalow by Frank Johnstone during his ownership, was enlarged and remodeled into the stately, imposing columned mansion of today by John Miller, who also laid out and planted the gardens, with the exceptions of a few palmettos and certain other shrubbery placed here by his predecessor.[75]

It was from the Miller ownership that the present owner, George L. Buist purchased the plantation. He and Mrs. Buist make frequent visits here from Charleston during the year.

Sampit River and Winyah Bay Addenda

Friendfield

Happily, Friendfield can be added to the list of Georgetown plantations that have remained relatively the same over the last quarter-century. The house and grounds continue to be quite lovely.

The plantation is also still owned by the Cheston family. Radcliffe Cheston, Jr., of Philadelphia, Pennsylvania, bought the property, consisting at the time of Friendfield and four other plantations, in 1930. The tract amounted to some 3,500 acres. Upon Mr. Cheston's death in 1968, this large estate passed to his three children, George Morris Cheston, Mrs. Sidney Cheston Devereux, both of Philadelphia, and Mrs. Frances Cheston Train of Bedford, New York. A few years ago, Mrs. Devereux died, and her interest is now held by her husband Antelo Devereux and her two sons, Antelo Devereux, Jr., and John Devereux. Members of the three Cheston branches spend time here periodically.

Maryville

The major development here in recent years has been the plantation's total transformation into the urban community of Maryville directly across the harbor from Georgetown. Not even the old, charmingly quaint plantation house remains.

Charlotte Ives Montgomery, daughter of Colonel and Mrs. Robert L. Montgomery of Mansfield plantation on Black River, purchased this portion of Maryville in 1944. The greater part of the plantation was already undergoing urbanization at this time. Charlotte Montgomery lived in the Maryville house, a dwelling dating back to 1811, until 1970 when she moved to a new residence nearby. In June of 1970, the 1811 house was dismantled and its timbers sold. In 1972 Mrs. Montgomery returned to Philadelphia, where she resided until her death.

This part of Maryville plantation land passed in 1973 from Charlotte Montgomery to Dr. Harry C. Tiller, Marvin M. Thomas and Arthur Thomas, all of Georgetown. Arthur Thomas died in 1984, leaving his interest in the property to his three children, Eric Thomas, Mrs. Sally Thomas Huntley and Mrs. Betsy Thomas Lovall. Marvin Thomas died in 1989. His heirs, Arthur Thomas's heirs and Dr. Tiller sold the property in 1991 to International Paper Company.

Belle Isle

Belle Isle in 1972 joined the ranks of local plantations being developed into resort residential communities. The Frank E. Johnstone family of Georgetown had owned Belle Isle since 1928 and had opened the gardens each year to the public. Upon Frank Johnstone's death in the 1960's, his son William A. Johnstone assumed management. It was under William that the plantation's development as a resort residential complex began and is continuing today. The complex consists of a yacht club, private homes and condominiums.

Still one of the loveliest, most impressive settings on Winyah Bay, Belle Isle even now retains the interesting breastworks of the Civil War Battery White overlooking the water and retains, also, the multitude of azaleas and camellias once featured in the gardens. This combination of history and beauty makes the place of special importance to Georgetown County.

Dover

Noted especially for its fine, old house, which was originally the main dwelling at Woodlawn plantation in Berkeley County and which was moved to its present site, Dover remains largely unchanged. Mrs. Henry M. Sage, who re-erected and restored the imposing Woodlawn house here in 1949, died on May 9, 1972. She left the property to her son Henry M. Sage, Jr.,who died in 1990. His four children, H.M. Sage, III, Linn Sage Jackson, Reginald D. Sage and Daniel C.T. Sage, now own Dover consisting of 213 acres.

Estherville

Estherville is unchanged as far as house and grounds are concerned, and this stability can be considered very fortunate. The plantation continues to be among the most beautifully planted and maintained places in Georgetown County. The banks of blooming azaleas here each spring present a scene of spectacular color.

There have been some changes in acreage and ownership, however. Mr. and Mrs. George L. Buist of Charleston bought the 3,000-acre property in 1943. Following Mr. Buist's death, Mrs. Buist sold the plantation in August of 1971 to Estherville Corporation composed of three sisters, Martha Parker Allison, Sarah Parker Lumpkin and Penelope Parker Peterson, all Georgetown natives. Sarah and her husband Dr. Robert L. Lumpkin made Estherville their home in March, 1972. In February, 1980, Estherville Corporation transferred the plantation to another family entity, the Parker Corporation, which, in turn, immediately transferred it to the present owner, Mrs. Lumpkin. Dr. and Mrs. Lumpkin continue to reside here. Since 1971, a substantial amount of the land has been sold. Estherville now consists of 1,000 acres. The plantation is still one of the most impressive settings on Winyah Bay.

A portion of former Estherville property is now being used for the commercial production of shrimp. A group of South Carolina investors formed Sea Fare, Incorporated, in 1987 and proceeded to flood a thirty-five-acre former rice field adjacent to the inland waterway for the purpose of raising shrimp on a large scale. Proximity to the mouth of Winyah Bay made the necessary flow of saltwater possible. The company plans to expand operations in the next few years to include two hundred acres of flooded area devoted to the production of shrimp. Already producing some eight million shrimp, Sea Fare believes its expansion will place the company among the largest aquaculture enterprises in South Carolina. The shrimp are to be sold throughout the United States. This relatively new use of old rice fields for aquaculture opens up all kinds of possibilities for local plantations.

NORTH SANTEE

Cat Island

SEPARATED FROM THE REST OF GEORGETOWN COUNTY by the Inland Waterway canal, the large area of land composing Cat Island includes four estates. The one among them called Cat Island has been owned for the greater part of its existence by the Lowndes family.

A small portion of the island originally was part of Winyah Barony.[1] Later, the whole island—which consists principally of swampy rice land—was held by Daniel LaRoche, an early settler in the Winyah country. LaRoche accumulated a large fortune and at his death left to be disposed of at sale some 10,000 acres of land on Black and Waccamaw rivers, the whole of Cat Island, eight lots in Georgetown, and 130 slaves, as well as cattle, horses, boats, and canoes.[2]

After LaRoche's time, the island became divided into various holdings. Thomas Lowndes, son of Rawlins Lowndes, acquired what is now Cat Island plantation, a tract located nearer to North Santee River than to Winyah Bay.[3] Rawlins Lowndes, who climaxed a life studded with other political honors by serving as Governor of South Carolina in 1778, came to South Carolina in 1730 and died in 1800.[4] His son Thomas, born in 1766, married Sarah, daughter of Jacob Bond I'On of Charleston,[5] and had a son, Richard Henry (born in 1815), who inherited Cat Island plantation.[6] Richard Henry made his home here in the shingle-sided residence that stands today[7]—a spacious home situated amidst live oaks and tall pines. The house is in a tranquil spot overlooking the rice fields, where in cold weather flights of wild ducks descend to seek refuge. For years the place was called Chat Isle,[8] but since Richard's ownership it has been known more commonly and simply as Cat Island.

The story is told that Thomas Lowndes, noting his son's love of horses, was worried about Richard Henry's future career. Thomas once came upon Richard Henry at the stable, trying out a horse, and remarked dryly, "Well, Richard, I suppose this is what you'll be all your life, a stable boy." The boy replied, "No, sir. I am going into the navy."

CAT ISLAND

(House no longer exists)

And into the navy he went in 1831, earning his commission in special schooling aboard ship and serving until 1842. Three years later he married Susan Middleton Parker and settled down to planting rice at Cat Island.[9] During the War Between the States, it is said, his rice-mill chimney here was used frequently for target practice by his former naval comrades aboard Federal gunboats prowling the coast. He gleefully assured some of them after the war that the chimney had never been struck.[10]

Richard's daughter Caroline married Dominic Lynch Pringle, who lived at Greenfield on Black River. His other daughter, Emily Rutledge, married Charles Petigru Allston, who purchased and lived at Windsor.

Richard Henry Lowndes' son, Richard I'On, eventually was given Cat Island plantation by his father, who died in 1905.[11] After Richard I'On's death, his heirs in 1919 sold Cat Island and adjoining Belle Isle, now known as Black Out, to B. Walker Cannon.[12] He, in 1920, sold both places to Thomas C. Gower of Greenville, South Carolina,[13] who, seven years later, conveyed this acreage to Mrs. William Gouverneur Ramsay of Wilmington, Delaware.[14] Mrs. Ramsay transferred Belle Isle to her son-in-law and daughter, Mr. and Mrs. William E. Phelps. She retained Cat Island, and upon her death in 1957, the place passed to her grandchildren, M. P. Ferris of Georgetown, Alfred D. Chandler III of Cambridge, Massachusetts, and W. Laird Stabler, Jr. of Wilmington, Delaware.

To reach any part of Cat Island one must cross the Inland Waterway canal by a hand-operated ferry—one of the few remaining in South Carolina. Until recent years, it was powered by strong muscles, a long pole, and the prayers of passengers; now, having surrendered to progress, the ferry shuttles back and forth under the comparative reliability of a gasoline motor—all very picturesque and time-consuming.

Black Out

BLACK OUT STEMS FROM LAND HELD BY RICHARD I'ON Lowndes, who was owner for some years of the adjoining Cat Island plantation. The property was originally named Belle Isle, but the present owners, Mr. and Mrs. William E. Phelps, call it Black Out—rather appropriately considering the plantation's isolation in the forests of Cat Island. The estate is almost five miles from the Inland Waterway ferry and faces in the general direction of the North Santee River.[15]

Belle Isle, as the tract used to be known, probably was part of the property given to Richard I'On Lowndes by his father, Richard Henry. It still belonged to Richard I'On at his death and was sold finally with Cat Island and other tracts to B. Walker Cannon in 1919.[16] Cannon then conveyed both Belle Isle and Cat Island in 1920 to Thomas C. Gower, of Greenville, South Carolina,[17] from whom the two estates passed to Mrs. William Gouverneur Ramsay of Wilmington, Delaware, in 1927.[18] Mrs. Ramsay continued to hold Cat Island but transferred Belle Isle to her son-in-law and daughter, Mr. and Mrs. Phelps.

The Phelpses changed the plantation's name from Belle Isle to Black Out and erected a low rambling-style, hunting lodge here for use during the winter hunting season, during which the whole region about Cat Island provides excellent duck-shooting.

Near Black Out are two other plantations—Humes and White Marsh, held now by Thomas A. Yawkey, owner of the Boston Red Sox baseball club, who owns South Island, where he has a winter home. Both of these places have a rice-planting history. Humes belonged at one time to Dr. Alexander Hume.[19] After the War Between the States, it was acquired by the Bennett family and sold by them in 1897 to Herman S. Cordes, who conveyed it in 1903 to the distinguished General E. P. Alexander.[20] White Marsh, once owned by the Maxwell family, was also purchased by the General,[21] whose principal holding at the time was South Island.

BLACK OUT

A native of Washington, Georgia, and a graduate of West Point, General Edward Porter Alexander had commanded the artillery in General Longstreet's corps of the Army of Northern Virginia. He was the officer who directed fire in the artillery duel at Gettysburg. During Grover Cleveland's last years, the President often hunted with General Alexander at South Island and Cat Island.[22]

Eventually the General sold Humes and White Marsh to J. H. Wheeler, from whom they passed to Thomas Yawkey.[23]

Annandale

(Millbrook)

ANNANDALE, SITE OF THE FIRST TIDE-OPERATED RICE mill ever constructed in the state, is also distinguished as the one-time property of the treasurer of the Confederate States of America.

Annandale, originally known as Millbrook, is traced to the ownership of Andrew Johnstone, only surviving son of Archibald Johnston—who spelled his name without the *e*. Andrew was the first in his family to own Estherville. In 1791 and 1792 when the plantation was called by its old name, Jonathan Lucas, inventive genius of the rice industry, constructed here the first tide mill for processing rice, and for years afterwards the mill operated regularly.[24] The plantation was destined to belong eventually to George Alfred Trenholm, Charleston merchant, who served a year as the Confederate secretary of the treasury, but through the pre-war years, it remained in the Johnstone line.

Andrew Johnstone married Sarah MacKewn in 1773 and made his home at Estherville.[25] Sometime before 1791 he acquired the estate then known as Millbrook.[26] Andrew left at his death a son, William, and a daughter, Esther. William inherited Millbrook, where he planted rice and operated the mill, and he also inherited Estherville, though Esther and her husband, Robert Withers, are thought to have had it at one time. Even so, the Withers couple apparently lived at Millbrook with the Johnstones from about 1812 to 1822 because, it is believed, the Estherville house had burned by this time.[27] Esther and her children perished in the 1822 hurricane that wrecked North Island.[28]

About 1830 Millbrook passed from William Johnstone to his eldest son, Andrew, who held it until 1863. While he was a young man, Andrew traveled abroad. Upon his return home he made two important changes at his plantation. Having met kinsmen in Scotland, he decided to rename his own establishment on North Santee for their home. Millbrook thus became Annandale. Andrew Johnstone also erected a fine house on the plantation, and the columned, stately mansion built here by him in 1833 in a grove of live oaks still stands as the main Annan-

ANNANDALE (MILLBROOK)

dale residence.[29] It is located about a quarter of a mile north of the site of the old Millbrook house that burned. Originally, there was a wing on the south side, but it was removed in 1924.[30]

The Johnstone reign at Annandale came to an end on August 8, 1863, when Andrew Johnstone conveyed the estate, consisting at that time of 2,290 acres, to George Alfred Trenholm. The property was then composed of five tracts—Millbrook, Annandale, Tidyman's Rush Land, Lynch's, and a part of Estherville—all united into one holding[31] and called Annandale. A year later Andrew Johnstone died.[32]

George Trenholm, born on February 25, 1807, in Charleston and married to Anna Helen Holmes of Johns Island in 1828, became one of his native city's most successful merchants. His experience in finance put him in a good position to help the Confederacy, and in July of 1864, he was called upon to serve as secretary of the treasury for the tottering Confederate States of America. Trenholm could not have come to the office at a worse time; he entered the government just in time to assure himself of punishment by Federal authorities. He remained in Richmond from the date he became treasurer until collapse of the government in April, 1865, and was arrested and imprisoned by United States officials in Charleston on June 13 of that year. On the eighteenth he was moved to Fort Pulaski, and on June 25 he was paroled and allowed to return to Charleston, where he was re-arrested on July 12 and again put into prison, until October 14.[33]

On December 27, 1865, Annandale was conveyed by Trenholm to his son-in-law, William Miles Hazzard,[34] who had married Emily St. Pierre Trenholm on January 21, 1864.[35] The property continued in the Hazzard family for a number of years. In 1898 it came into the possession of the Santee Rice Company, which conveyed it to Wayne Darlington in 1911.[36]

Darlington, related by marriage to Annandale's old ruling family, the Johnstones, kept the plantation until 1919, when it was purchased from him by Charles M. Reeves of Sanford, North Carolina. Transferred in this sale, too, were Hopeland and Hopewell. Charles Reeves then sold the three places, Annandale, Hopeland, and Hopewell, to Richard E. Reeves of Summit, New Jersey, in 1924, two years before the latter's death.[37] Annandale is now owned by his heirs: a son, C. E. Graham Reeves; a daughter, Mrs. Susan Reeves Deland; and the widow, Mrs. Susie Reeves Early.

Under the Reeves family, Annandale has taken on new life. Richard E. Reeves in 1925 restored the fine old residence, which had been unoccupied since the Hazzards left the plantation. He also conceived a new use for Annandale's abandoned rice fields. Rice planting on a large scale had continued until about 1908, but thereafter, despite Wayne Darlington's efforts to revive production, had dwindled and finally ceased. Mr. Reeves believed that the old fields would make excellent grazing land, and at the time of his death was planning to use them for cattle raising.

Only recently has Richard Reeves' dream come true. In 1949 C. E. Graham Reeves and his sister, Mrs. Deland, undertook an extensive clearing and reclamation project in order to put the plantation in the cattle business. To date about 500 acres of upland have been cleared and seeded to permanent pasture, and approximately 500 acres of rice field land diked to keep out salt water and to make them also suitable for grazing land. The project has been successful, and the plantation herd now numbers more than four hundred head of cattle.[38]

Each winter the Reeves family comes to Annandale for a stay, to enjoy the camellias and azaleas in bloom and to take in the fine duck-hunting to be had here.

Ravenel

(Rochelle)

OWNED FOR MANY YEARS BY THE RAVENEL FAMILY, FOR whom it is named, Ravenel plantation on the North Santee River is now the possession of the Manigaults of Charleston.

The estate has occasionally been called Marsh plantation[39]—the dwelling being situated on a small piece of high ground practically surrounded by marsh and swamp; but Ravenel is the name more commonly used. Long before the War Between the States, this property belonged to the well-known Ravenel family of South Carolina, and their ownership continued into post-war years.[40]

Alfred G. Trenholm, son of George A. Trenholm, the Confederate States' last secretary of the treasury and one-time owner of Annandale, acquired Ravenel plantation sometime after the war, and it was part of his estate when he died intestate in 1898. His heirs sold the place the following year to Alfred's sister, Emily St. Pierre Hazzard, wife of William Miles Hazzard, whom the court had appointed as Alfred Trenholm's administrator. Emily Hazzard then sold in 1899 to Arthur M. Manigault,[41] son of General Arthur M. Manigault, the officer who had charge of organizing Winyah Bay and Santee River defenses at the beginning of the War Between the States.

Manigault also purchased adjoining Lynchfield plantation,[42] granted originally to Thomas Lynch and held by descendants of the Lynch family until its acquisition by Manigault. Since this sale Lynchfield and Ravenel have changed hands as a unit.[43] Rice was grown commercially until 1907 on this combined tract, which the Manigaults called Rochelle.

By the time he became owner of Ravenel, Arthur Manigault already had spread his business interests in another direction. In 1895 he had bought controlling interest in a Charleston newspaper, *The Charleston Evening Post.* His son, Robert Smith, became president and publisher of *The News and Courier* after its purchase by *The Charleston Evening Post* and others in 1926.[44]

RAVENEL (ROCHELLE)

Ravenel

Ravenel plantation was purchased in 1919 by L. K. Estridge of North, South Carolina, in connection with a hunting club.[45] John Miller, owner of Estherville, not long afterwards acquired it for the Winyah Gun Club, and the late Robert Smith Manigault secured the property from this ownership, thus bringing Ravenel back into his family.[46]

Edward Manigault, who succeeded his deceased brother, Robert, as publisher of the Charleston newspapers, also followed him as owner of the plantation and is responsible for outstanding work in restoration and improvement recently achieved here. In 1957 he remodeled the old house built by his father at the turn of the century[47] and has made of it one of the loveliest houses to be found among all the Georgetown plantations. In addition, various shrubs have been planted over re-cleared grounds to add to the general beautification of the place.

Church of the Messiah, North Santee

(Church of the Messiah, Maryville)

LIKE SCORES OF OTHER RURAL CHURCHES OF THE PLANTATION country, the Protestant Episcopal Church of the Messiah, North Santee, died with the end of rice planting in South Carolina. No service was conducted in this picturesque little building from about 1929[48] until the church's removal in the spring of 1955 to Maryville, where it now is the thriving mission of Church of the Messiah.

A general exodus from the coastal plantations took place in the early 1900's as planters sought more profitable means of making their living, and with the departure of its congregation, the Church of the Messiah had little choice but to close its doors. Rebirth took place in early 1955 as the Rev. A. Nelson Daunt, new rector of Prince George, Winyah, and All Saints, Waccamaw, churches, organized an Episcopal mission in the small town of Maryville across the Sampit River from Georgetown. The Diocese of South Carolina gave permission for the old Santee structure to be moved and to become the church of the Maryville congregation. Then individually and collectively, members of the other two Episcopal parishes presented gifts to the revived Church of the Messiah.

The late Mrs. Charles Weston Rosa gave a silver paten in memory of her small son, Charles Weston, Jr., and a pair of brass Communion candle sticks in memory of a former rector, Rev. Dr. J. C. Waring. A silver bread box was given by Mrs. Frederick Wentworth Ford in memory of her parents, William J. and Margaret Deas Lucas; and her daughters, Mrs. C. V. Steinhart and Mrs. Dwight D. Morris, presented a hymn board and brass alms basin, respectively, in memory of their father, Frederick W. Ford. The brass altar cross was a gift from T. Cordes Lucas in memory of his wife, Alice Vaughan Lucas, and a pair of brass candelabra was presented by Mrs. Arthur Herbert Lachicotte. Robert V. Read of Columbia, South Carolina, and his sister, Mrs. Alan Wood of Philadelphia, gave a brass processional cross and a pair of English glass-and-silver cruets in memory of their parents, Mr. and Mrs. John Harleston Read III; and Mrs. J. C. Waring, the

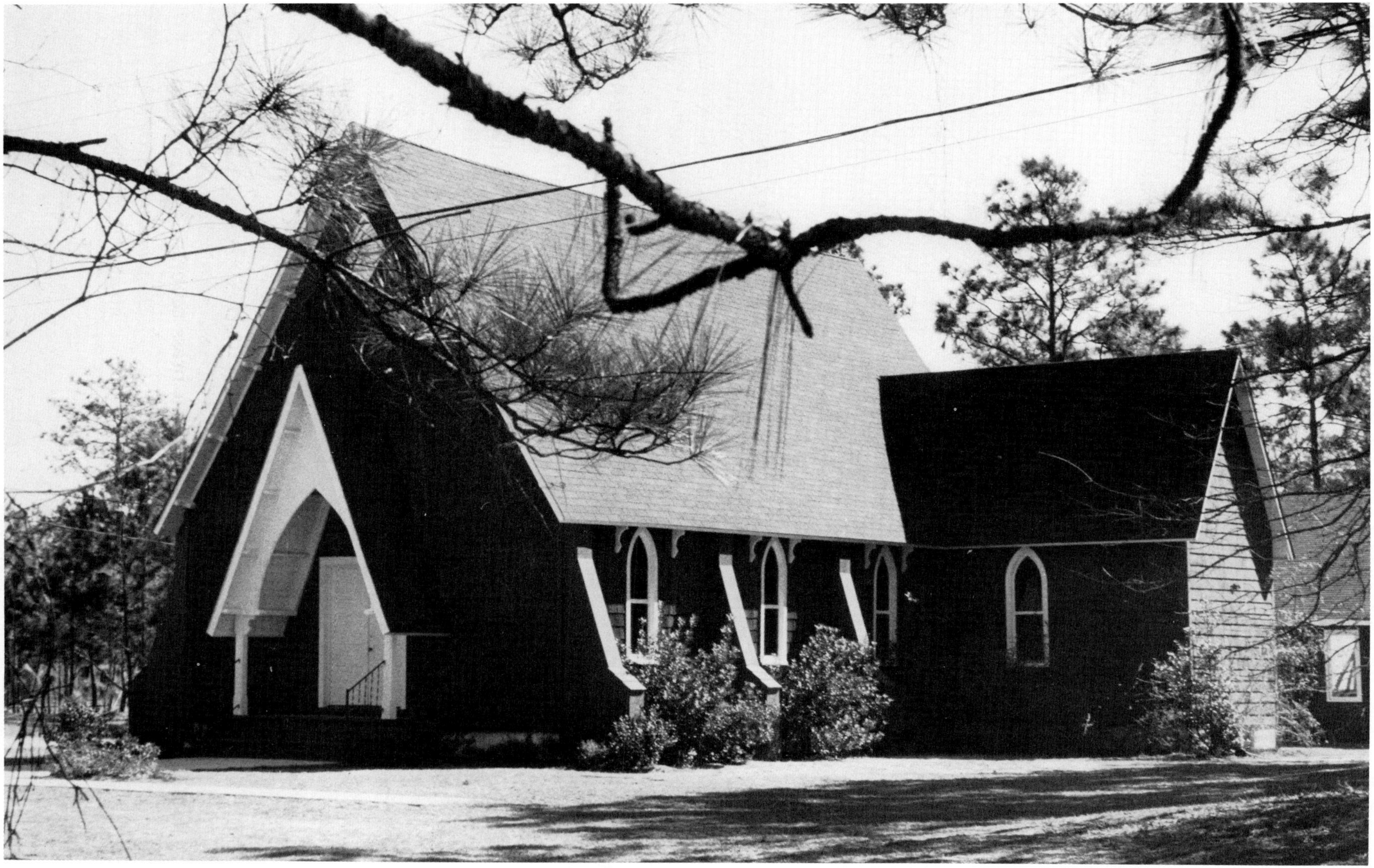

CHURCH OF THE MESSIAH, NORTH SANTEE
(CHURCH OF THE MESSIAH, MARYVILLE)

(Renovated building not shown in previous printings)

widow of Dr. Waring, presented a chancel hymn book and prayer book. A baptismal bowl in memory of Frederick Wentworth Ford, Jr., was the gift of Mrs. William G. Ramsay of Wilmington, Delaware; and the bishop's chair was given by Mrs. Theodore K. Selkirk of Cincinnati and her brothers, William C. and H. Fraser Johnstone in memory of their parents, Mr. and Mrs. William H. Johnstone.

The establishment of the first Church of the Messiah began in 1804 with the erection of a church building at North Santee ferry. A group of residents in the Santee region had asked the General Assembly of South Carolina to incorporate them into a body politic under the name The North Santee Church in the Parish of Winyaw. Fourteen petitioners were listed: James Hamilton, John Hume, Elias Horry, Isaac Mazyck, Paul Mazyck, Thomas Horry, Esther Lynch, John Drayton, Seaman Deas, William Johnston, L. D. Wigfall, Benjamin Smith; Philip Tidyman, and Henry Deas.[49]

The new church existed in this status and was known only as the Church on North Santee until 1838.[50] In 1819 the congregation was being served by a lay reader.[51] C. C. Pinckney was the last minister the church had under its old organization. In 1842 the church was re-organized under the name Church of the Messiah, North Santee, and separated from the parish of Prince George. The church was associated more closely with St. James, Santee, and the two were often served by the same priest.[52]

Sometime before the War Between the States, the building at North Santee ferry was dismantled, moved, and re-erected farther down the river. Described as much larger than the present building, it was designed in the shape of a cross.[53] Church of the Messiah at one time had a summer chapel on South Island.[54]

Thomas J. Girardeau was rector in 1855, and in the years from 1872 to 1874, the church was served by W. H. Johnson. He went occasionally from his charge to All Saints Parish, Waccamaw, to visit and to conduct services.[55] W. O. Prentiss was serving as rector in 1881, and in that year the church included thirteen families, constituting twenty-two communicants and fifty "whole number of souls."[56]

On December 2, 1886, the old church burned. It was replaced with the present, smaller building in 1901, and the consecration was held in April of the same year by the Rt. Rev. Ellison Capers, D.D., Bishop of the Diocese of South Carolina.[57] Of Gothic design in exterior and interior architecture and in furniture, this little church is one of the quaint old buildings in the low country. A particularly intriguing feature of the structure before its re-erection at Maryville was the roof where various patterns had been worked out with the shingles.

Church of the Messiah took fresh life under the rectorship of J. C. Waring in 1900, but the changing economy in those last years of rice planting shortly doomed it to many years of inactivity. Robert Wilson of Charleston was the last minister to serve, and the church doors closed in 1909. The property, by canon law, reverted to the diocese in 1920.[58] In 1929 Henry DeSaussure Bull, rector of Prince George, Winyah, conducted the last service held in the church on its North Santee site. Some of the old members and various interested persons gathered for the occasion.[59]

Today this small Gothic church rests on a new site in Maryville, where the outlook for future growth is bright. Undoubtedly, the Church of the Messiah is entering a new era of prosperity.

Hide Out

(Wicklow Hall)

HIDE OUT IS THE UPLAND PORTION OF WICKLOW PLANtation, most of which is a part of Kinloch property now. Since 1927 it has been in the possession of T. Cordes Lucas, who recently gave the house and surrounding premises the name Hide Out.[60]

The rice fields of Wicklow, some 380 to 480 acres, are on the delta between the North and South Santee rivers, but the house is situated on a separate strip of land altogether.[61] It has been known by several names—The Strip, Duck Point Dwelling, Lowndesville, and Wicklow Hall. This residence was built about 1850 by carpenters belonging to Stephen D. Doar, the grandfather of Cordes Lucas, for Rawlins Lowndes, Wicklow's owner of that time. Lowndes usually spent part of each year in New York City but wanted a dwelling near his rice fields when he came south.[62]

Wicklow, both rice fields and house, remained in the Lowndes family until 1899, when the heirs of Rawlins Lowndes sold the plantation to S. M. Ward and Company, a rice-planting firm owned by S. M. Ward, St. Julian M. Lachicotte, and A. A. Springs;[63] and Wicklow was included in the vast acreage on North Santee and the delta that this company conveyed in 1912 to the Kinloch Gun Club. The land involved in that transaction went far toward filling out the present boundaries of the Kinloch estate.[64]

In 1927 the Wicklow house and grounds were deeded to T. Cordes Lucas by the Kinloch Gun Club.[65] Beginning in 1918, Cordes Lucas served thirty-six years as manager of the large Kinloch plantation. Only recently did he retire from this position. A descendant of Jonathan Lucas, the inventor whose improvements in rice-milling spurred the rice industry to new heights, Cordes Lucas has spent practically his entire life on the Santee. Born at Harrietta plantation, his mother's old home on the South Santee, he later lived at the family plantation, Hopsewee, on North Santee. With the exception of sixteen years in Philadelphia, Mr. Lucas has spent all his life in this section.[66] He is

HIDE OUT (WICKLOW HALL)

(Renovated house exterior not shown in previous printings)

so much a part of the countryside where he makes his home and so deeply interested in its history that he is considered an authority on the region. The great Santee could not wish for a more devoted, eloquent spokesman.

Courtesy of Mr. John Lofton

OLD MILLDAM RICE MILL AT KINLOCH

Kinloch

COMPOSED OF THIRTEEN ONCE INDEPENDENT PLANTATIONS, Kinloch now includes some 9,000 acres and is one of the largest single holdings on the North Santee. This vast aggregation of property is named for nearby Kinloch Creek, which, in turn, received its name from Francis Kinloch, an early owner of Rice Hope.[67]

Kinloch house stands on the site of the old Milldam plantation,[68] owned at one time by Thomas Horry.[69] The original dwelling burned in 1879 and was replaced in 1923 by the present mammoth structure containing eight bedrooms, large living and dining rooms, a gun room, office, kitchen, and pantry. The immense proportions of the building are due, probably, to the date of construction, which took place during the existence of the thirty-five-member Kinloch Gun Club.

Present owners of the estate are Delaware industrialists, Eugene E. du Pont and a cousin, Eugene du Pont, whose father along with Eugene E. gained ownership by taking over the interests of the Kinloch Gun Club. The old places they own today under the name of Kinloch are, besides Milldam, Newland, Campmain, Richfield, Pleasant Meadows, Crow Island, Wicklow, Tranquility, Motfield, Bear Hill, Pine Grove, Jutland, and Doar Point. Some of these tracts are on the banks of the North Santee; others are on the delta between the North and South Santee rivers.[70]

Milldam, which contains the principal buildings connected with the operation of Kinloch, is traced to the 1700's and to the ownership of Thomas Horry, youngest son of Elias Horry II, who held part of Winyah Barony for a time. Elias also possessed considerable land on North Santee, some of which, it is concluded, must have passed to Thomas. In addition to Milldam, Thomas owned the plantations of Campmain and Newland, which were in the same area. Married in 1772, Thomas had a son, Elias IV, who eventually inherited his father's three plantations; and upon the death of Elias IV this property passed to his three sons, Thomas, Edward, and Paul, and to his daughter,

KINLOCH

Julia, the wife of Dr. Benjamin Trapier. The lands remained in the Horry family until after the War Between the States.[71]

In 1885 Milldam and Newland were combined into one place and were purchased by Arthur M. Manigault, son of General Arthur M. Manigault. Six years later Manigault sold the Milldam–Newland holding to S. M. Ward and Company.[72] During S. M. Ward and Company's ownership the Kinloch property expanded to almost its present acreage. When the company sold its Santee holdings in 1912 to the Kinloch Gun Club of Wilmington, Delaware, fourteen tracts changed hands.[73] Among them were the following plantations: Richfield, at one time the property of William Bull Pringle;[74] Bear Hill, owned in 1836 by Henry Deas, and later by Robert Hume and William Bull Pringle; Pleasant Meadows, in the early years held by William Matthews, who devised it to his daughter, Susan B. Hunt, the land passing in 1854 from William M. Hunt to William Bull Pringle; Milldam and Newland, one plantation at the time of Arthur M. Manigault's purchase in 1885;[75] White Oak, owned by General Arthur Manigault, afterwards by his son,[76] and sold in 1927 by the Kinloch Gun Club to William N. Beach of Rice Hope;[77] Campmain, an old Horry place, owned by the Frost family when sold to S. M. Ward and Company;[78] Pine Grove,[79] once a holding of the Trenholm family; Wicklow, for many years owned by the Lowndes family and now owned by T. Cordes Lucas.[80] Other plantations attached to Kinloch include Doar Point, obviously once owned by the Doar family; Crow Island, consisting really of two plantations, one owned by the Lucases and the other by the Deas family; Motfield and Tranquility, both former Rutledge places, and Jutland, a Lucas plantation.[81]

Kinloch is situated in splendid wild duck and deer country, and each winter sportsmen come here to enjoy the excellent hunting of the Santee region.

WOODSIDE

(New house not in previous printings)

Woodside

ONCE A PART OF RICE HOPE PLANTATION, OLD HOME OF the Kinloch and Lucas families, Woodside came into its own more than one hundred years ago. Its house is now one of the oldest in the North Santee rice-planting region.

From 1905 to 1956 the estate of Russell Middleton Doar, Woodside was inherited by him from his parents,[82] who purchased it from Edward Horry,[83] son of Elias Horry IV, owner of Milldam, Newland and Campmain—all now part of Kinloch. In 1956 Woodside, too, became a portion of Kinloch through its sale by the Doars to the Kinloch ownership.

The plantation came into being in the 1840's, when 411 acres were cut off from Rice Hope by Simons Lucas for his brother Henry. The land was deeded to Henry Lucas in two parcels, the first in 1843 and the second in 1846.[84] This acreage was on the wooded side of Rice Hope—hence the name Woodside. When Henry Lucas came into ownership, he erected on his property the house that stands today, intending it to look as much as possible like the old family home in England.[85] The two-story structure contains ten rooms and is supported in front by quaint hand-hewn pillars.[86] Upon leaving Santee to make his home in Georgia, Lucas sold the plantation to Edward S. Horry,[87] in whose hands it remained until 1879 when it was sold to Thomas Watts Doar.[88]

More fortunate than many other places, which were destroyed completely, Woodside house came through the War Between the States with only scarred and split doors. Instead of burning the dwelling, crewmen of Federal gunboats preying up and down the North Santee, satisfied themselves with beating open doors and breaking into the house. The front and hall doors still bear the marks inflicted by the gun butts of Yankee raiders.

Russell Doar died in 1960. Until recently he and Mrs. Doar had made their home at Woodside.

Rice Hope

A NOTABLE DISTINCTION HELD BY RICE HOPE PLANTATION is its early nineteenth-century ownership by the great rice-mill builder, Jonathan Lucas, Jr., upon whom the British government bestowed knighthood for the erection of rice mills in Great Britain.[89] Following in the footsteps of his father, who invented both the tide- and steam-mill processes for milling rice,[90] the younger Lucas improved on the elder Jonathan's basic designs.[91] Together father and son did more than any other men to boost the rice industry to its peak during the years just preceding the War Between the States.

Jonathan Lucas, Jr., acquired Rice Hope sometime before he returned to his native England in 1822 to construct rice mills there. Ten years later he died in Great Britain, but Rice Hope plantation remained in his family until 1926, when it was purchased by the late William N. Beach of Long Island, New York. Mr. Beach died on May 5, 1955, and a large portion of this property was sold off the next year.

Rice Hope's history goes back to the mid-1700's, when it was owned by the Kinloch family. The land was first the home of Francis Kinloch, only surviving son of James Kinloch of St. James Parish, Santee. Francis, born in 1720, inherited large holdings from his father and went on to establish other plantations, becoming a highly successful rice planter.[92] He was active politically and served from St. James Parish in the Commons House of Assembly from 1751 to 1754. In 1751 he married Ann Isabella Cleland, coming through this marriage into possession of Weehaw and Kensington plantations on Black River. Francis and Ann had three sons, James, Francis, and Cleland, and a daughter, Mary Esther. To all four children, Francis left his plantations of Rice Hope, New Gilmerton, Weehaw, and Kensington. Under age at the time of their father's death at Rice Hope in 1767, the children were placed under the guardianship of Thomas Boone, one-time Governor of South Carolina.[93]

An outstanding figure of the state in later years, young Francis Kinloch, before settling down—if really he ever did—received his final education in Europe. Schooled first by a private tutor, he was sent to England in 1768 to enter Eton College, from which he was graduated in 1774. He later traveled on the continent for three years, studying in Rome and Geneva with the idea of entering the diplomatic service. In Switzerland he was a fellow student of another South Carolinian, Henry Laurens. Kinloch finally returned to America, after spending a year in Scotland with relatives, and immediately entered the American army as a lieutenant. He served in a military capacity from 1778 until 1781, ultimately attaining the rank of captain. Wounded in one battle, he saved his bullet-pierced coat and later had the hole embroidered. In after years he wore the coat with immense pride. He was twice captured by the British—first at the fall of Charleston, and after securing release, in Virginia, where he had rejoined the American forces.

Although the recipient of many public honors, Kinloch is supposed to have been rather indifferent to office-holding. He served as a member of the Continental Congress and of the South Carolina General Assembly, and in 1788 he was a delegate to the state convention that ratified the United States Constitution. In 1790 he was a member of the state constitutional convention. After that year he refused to accept any other positions.

Two family trips to Europe, one in 1790 and the other in 1802 for four years so depleted Francis Kinloch's fortunes that he was forced to sell all of his property except Kensington, where he lived until it was sold in 1824. He died in Charleston in 1826. Rice Hope plantation had passed out of Kinloch possession much earlier, however, having been conveyed on July 2, 1789 to Edward Crook and George Lockey.[94]

Loss of some of the Lucas family papers makes it impossible to trace the exact length of their ownership and the date of the plantation's purchase by Jonathan Lucas, Jr. The elder Lucas was born at Cumberland, England, in 1754 and married Mary Cook there in 1774.[95] He was a skilled millwright when he arrived in South Carolina soon after the Revolution. In 1787 Jonathan Lucas erected the first water mill of his construction at Peach Island plantation on Santee. Later he built a water mill on Winyah Bay for General Peter Horry and one at Fairfield on the Waccamaw for Colonel William Alston. The first tide mill was erected by him in 1791 and 1792 for Andrew Johnstone at Millbrook, now part of Annandale. About a year or two afterwards,

RICE HOPE

Lucas built an improved tide mill, equipped with rolling screens, elevators, and packers, at Henry Laurens' Mepkin plantation on Cooper River. Later mills were "improvements in construction but not in plan," it is said.[96]

Together, Jonathan Lucas and his son, Jonathan, Jr., constructed in 1801 on the Cooper River the first toll mill for cleaning rice, and the first brushing screen ever in use was placed in this mill during 1803. In the early years of the nineteenth century, Jonathan Lucas erected a tide mill on the Ashley River near the West Point mill property. "This mill attracted a large business and made Charleston the center of the rice-milling industry." The use of steam power in rice milling was employed for the first time in a steam rice-mill constructed by Lucas in 1817 in Charleston.[97] The elder Lucas died at Charleston in 1821.[98] A year later Jonathan, Jr., went to Great Britain,[99] where he died in 1832, leaving a large estate there, in South Carolina, and in India.[100]

Simons Lucas, Jonathan, Jr.'s son, was living at Rice Hope in 1836, and after the War Between the States, his eldest son, William Johnstone Lucas, became owner. William Lucas sold the plantation in 1896 to his son-in-law, Frederick Wentworth Ford, who married Mary Ashe Lucas. Ford planted rice here until the summer of 1908, when a severe freshet destroyed all rice crops on North and South Santee rivers.[101] In 1926 he sold Rice Hope, consisting then of 1,113 acres, to William N. Beach, president of the Sterling National Bank and Trust Company of New York City.[102]

Upon acquisition of the property, William Beach made an addition to the east wing of the plantation house and remodeled the entrances, otherwise making necessary repairs without changing the architecture.[103] Though an old structure, this house is not the original dwelling. The first house, which is understood to have been located a little southeast of the present one, burned in the early years of the nineteenth century.[104]

Rice Hope is a lovely plantation; even before the War Between the States, it was noted for the gardens and greenhouse here. Roses, particularly preferred by the master and mistress of the estate, were planted over the entire grounds, and gardenias and weeping willow trees bordered a nearby lake, along which to this day the charming tree-shaded entrance drive makes it approach.

An event adding further luster to this Santee plantation's rich history is the fact that the Marquis de Lafayette stopped at Rice Hope during the Revolution, as he was journeying to Charleston after his landing on North Island. At that time the public road passed through the plantation and joined the King's Highway to the west, the route Lafayette is supposed to have taken.[105]

Rice Hope became a tremendously large holding of land. Absorbed within its boundaries are the plantations of White Oak, an adjoining tract of 1,035 acres,[106] once owned by General Arthur Manigault, then by his son and later by S. M. Ward and Company, which conveyed it to the Kinloch Gun Club, from which William Beach obtained it; Legrange, originally part of Hopsewee and sliced off in 1825 for John Hume, Jr., and his brother William—passing from them to another ownership and on to Dr. Samuel Cordes and his sons, and next to the Ladson family and finally to W. N. Beach; Fawn Hill, also a Ladson place at one time;[107] Prospect Hill and the large part of Dover, purchased in 1935 by Beach, who obtained at the same time Retreat plantation and parts of Belle Isle and Mount Hope.[108]

Following William Beach's death, the sale of various Rice Hope acreages began, one of the largest transactions having been the purchase of some 15 tracts in August, 1956 by Koppers Company, Incorporated. Mrs. Beach died September 10, 1960. The remaining portions of the plantation, including the house and surrounding grounds, then were obtained from the Beach estate in the autumn of 1961 by Williams Furniture Corporation of Sumter, South Carolina. Such is the present status of this old Santee plantation.

Hopsewee

ILLUSTRIOUS HOPSEWEE ON NORTH SANTEE HAS BEEN from its very founding one of the nation's historic homes. It was the plantation of Thomas Lynch, a South Carolina delegate to the Continental Congress, and the birthplace of his son, Thomas, Jr., who signed the Declaration of Independence.

The tract of land comprising Hopsewee was granted to the elder Thomas Lynch in the first half of the eighteenth century—possibly included in the grant he is known to have obtained on August 28, 1733, for 4,500 acres lying mainly south of Winyah Barony and including swamp land along the Santee.[109]

Between 1735 and 1740, Lynch constructed here the plantation home that still stands.[110] Among the earliest houses built in Georgetown County, it is the dwelling in which Thomas Lynch, Jr., was born on August 5, 1749.[111] The younger Lynch lived at Hopsewee until his father sold the place in 1762[112] to Robert Hume, a Goose Creek planter, son of Peter Hume, who settled in the Goose Creek section about 1729.[113]

Educated at Eton and the University of Cambridge, Thomas, Jr., studied law at the Temple in London. In 1772 he returned to South Carolina[114] and was married the same year to Elizabeth Shubrick.[115] Later he served in the First and Second Provincial Congresses and as a captain in the first regiment of provincial regulars recruited by the state as hostilities broke out in 1775. Young Lynch resigned his army commission the following year, when he was sent to the Continental Congress. The elder Thomas also was a delegate to this Congress but did not become a signer of the Declaration of Independence. He became ill during the journey to Philadelphia and died in Annapolis, Maryland. Thomas Lynch, Jr., falling into increasingly poor health, made one of his last public acts the service on a committee in 1776 to draft a constitution for South Carolina. In a final, desperate effort to prolong his life, he sailed in late 1779 for Saint Eustatius, West Indies, where he hoped to secure passage on a neutral vessel to France.

HOPSEWEE

The ship left the Indies and was never heard from again. It is supposed to have been lost in a storm.[116]

Robert Hume, who purchased Hopsewee from the elder Thomas Lynch in 1762, had two sons, Alexander and John. The elder brother, Alexander, was killed in the siege of Savannah in 1779. John succeeded his father as master of Hopsewee, making his winter home at the plantation and his summer home in Charleston. He amassed, in the meanwhile, a fortune through his rice-planting operations.[117]

John Hume was offered the Scottish Earldom of Marchmont, left vacant at the passing of a Hume relative, but he refused, saying he would rather be "earl of marsh mud" in South Carolina than Earl of Marchmont in Scotland. Well he might, considering the handsome returns in rice money he was getting from the muddy lands of Santee—infinitely more than could ever have been gotten from the deteriorated estate of his noble kinsman. It appears, in fact, that all the earldom offered was the title; everything else connected with the estate was in need of repair.[118] John Hume contentedly lived out the rest of his life in South Carolina and at his death in 1841 left seven or eight children. A daughter, Charlotte, married William Lucas, son of Jonathan Lucas, well-known inventor of improvements in milling rice. A son born from this marriage, John Hume Lucas, inherited his grandfather's plantation of Hopsewee.[119]

The new master of the plantation married his first cousin, Eleanor Ball Simons, and established his home here in 1846 after the house had been closed for many years. The front veranda, which had fallen into decay, was replaced about this time with the present double porches. John Hume Lucas died at the age of thirty-one in August, 1853, leaving a widow and three small children. Mrs. Lucas afterwards made her home permanently in Charleston and came to Hopsewee only in the holiday seasons of December and April. The rice planting was carried on by an agent. In May, 1862, she left Hopsewee for the last time, as homes along the river were being evacuated. The Hopsewee slaves were rented out to neighboring planters, and rice was never planted here again.

During the War Between the States, the handsomely furnished house was raided by the Federals, who appropriated all they wanted and divided the rest among the Negroes. Mrs. Lucas' eldest son, William, was able to rescue only a few pieces of furniture. In 1900 William Lucas brought his family to Hopsewee to live, and they resided here until his death in 1914. He died at Harrietta plantation on the South Santee, his wife's old home, owned at this time by his brother-in-law. Following William's death, the house at Hopsewee was closed once again for a time.[120]

In 1945 the Lucas family sold the plantation, consisting then of 781 acres of highland and additional rice land to A. A. Donham,[121] purchasing agent of the International Paper Company. Two years later the present owner, Mrs. Reading Wilkinson, and her husband, the late Colonel Wilkinson, acquired that portion containing the residence.[122] The Wilkinsons had owned Maryville plantation on Winyah Bay before selling it in 1944 to Mrs. Charlotte Ives Montgomery. Colonel Wilkinson was the retired district army engineer at Charleston. He died at Hopsewee in March, 1954.

One of the most interesting of the old plantation homes, Hopsewee has in each room—including the kitchen—Adams candlelight molding as part of the interior finishing. The four rooms on each of the two main floors open into center halls that extend the entire length of the dwelling. The house overlooks the North Santee River.

At the rear of the old house is a large circle of camellias planted about 1846 by John Hume Lucas.[123] The shrubs have now reached a magnificent height. The circle apparently formed part of a design for the plantation's entrance laid out many years ago. The intention seems to have been for the driveway to stretch for some distance between rows of live-oaks and to end at the camellia circle.

Mrs. Wilkinson continues to live at her celebrated estate of Hopsewee.

OUTDOOR KITCHEN AT HOPSEWEE

North Santee Addenda

Cat Island

The loss of the old Cat Island house is probably the biggest change to occur at this plantation during the last twenty years. The pre-Civil War dwelling was dismantled and its material sold to various individuals in 1972 when efforts to make needed repairs failed.

Some changes in ownership also have taken place. Mrs. William Gouverneur Ramsay of Wilmington, Delaware, bought the property in 1927 and left it upon her death in 1957 to her grandchildren, M.P. Ferris of Georgetown, Alfred D. Chandler, III, of Cambridge, Massachusetts, and W. Laird Stabler, Jr., of Wilmington, Delaware, all of whom received equal shares. Sophie Chandler Consagna, Alfred Chandler's sister, later obtained half of Alfred's share. Then on October 2, 1984, she deeded her portion, a one sixth undivided interest in the 1,571.1-acre plantation to The Nature Conservancy. This acreage is now administered by the Tom Yawkey Wildlife Center headquartered on adjacent South Island. Alfred D. Chandler, III, still owns the remaining half of his share in Cat Island plantation. M.P. Ferris and W. Laird Stabler, Jr., continue to hold their original shares.

The property today is largely woodland and former rice fields. The owners use the land mainly for hunting. There is considerable timber here, but inadequate transportation from the island across the inland waterway so far has been an obstacle to marketing it. The house at Cat Island has not been replaced.

Black Out

The lodge and grounds at Black Out have remained about the same through the years. The changes have been in ownership and acreage. The present owner, Wilhemina Stabler Bergland, inherited some 300 or more acres and the house from her aunt, Mrs. William E. Phelps, upon the latter's death on November 19, 1977. Another portion of the property, a tract containing 376.5 acres, was left by Mrs. Phelps to Mr. and Mrs. George B. Daniels of Charleston. They, in turn, sold this part on June 14, 1982, to JLH Trust, an entity representing Mrs. Thomas Yawkey. This trust then deeded the tract to The Nature Conservancy on February 22, 1983, and on the same day The Nature Conservancy transferred it to the South Carolina Wildlife and Marine Resources Department, which operates the Tom Yawkey Wildlife Center on South Island. The tract, therefore, is now part of the Yawkey Wildlife Center, which owns some 5,000 acres on Cat Island at this time.

Wilhemina Bergland and husband Charles Bergland come each winter from their home in Connecticut for a stay at Black Out.

Annandale (Millbrook)

Two major land divisions in the 1980's have brought about a significant change in the structure of this plantation. In 1982, C.E. Graham Reeves, who had bought his sister's interest in Annandale in 1968, sold 1,200 acres and the house to Peter Manigault, owner of nearby Rochelle plantation. Mr. Manigault, chairman of the board of the Evening Post Publishing Company, publisher of both Charleston newspapers, then sold this tract to White Oak Forestry Corporation, a subsidiary of the publishing company. White Oak in 1986 split off the Annandale house and the surrounding twenty-eight acres and sold them to Louis N. Dunklin of Dallas, Texas. On July 29, 1988, Mr. Dunklin conveyed the house and same acreage to the present owner, S. Alexander Gregg of Hilton Head Island, South Carolina. Mr. Gregg traces his family connections to the Johnstones, early owners of the place.

This section, containing the house built in 1833, has been returned to its original name of *Millbrook*. Graham Reeves has retained exclusive use of the name *Annandale* for the some 4,000 acres he still owns. In effect, he still possesses approximately four-fifths of the plantation.

Annandale, under the Reeves ownership, has continued to be a working plantation. For many years cattle were raised here. Now, soy beans and corn are grown on the uplands, and shrimp and crabs are raised on the wetlands. This tract remains one of the larger plantations in the county.

Annandale (Millbrook) was placed on the National Register of Historic Places in 1973.

Ravenel
(Rochelle)

Rochelle, as the Manigaults prefer to call the plantation, has continued basically unchanged since remodeling of the house in 1957 by Edward Manigault, the owner at that time.

The principal change has been in ownership. Edward's son, Peter, inherited the property upon his father's death on May 26, 1983, and still owns it. He comes frequently to Rochelle from his home in Charleston.

Peter Manigault also has succeeded his father as Chairman of the Board of the Evening Post Publishing Company, which still publishes the two Charleston newspapers, *The News and Courier* and *The Evening Post*.

In recent years the Evening Post Publishing Company has accumulated through subsidiaries tract after tract in the North Santee area. Today, the company controls large land holdings in this part of Georgetown County.

Church of the Messiah, North Santee
(Church of the Messiah, Maryville)

Continuing its history of collapses and revivals, Church of the Messiah is at the present time happily in a period of revival. The present building was dismantled and re-erected in 1955 on Poplar Street in the Maryville section of Georgetown. Still under the Episcopal Diocese of South Carolina, the building became the site of a new Episcopal congregation and was served by the Reverend A. Nelson Daunt, rector of Prince George, Winyah, Church, Georgetown. Despite the hopeful beginning, the congregation dwindled, and in 1961 the doors closed. Two years later a different group, the Pentecostal Holiness Church, acquired the building and established a congregation. The Pentecostals flourished so vigorously that by 1979 they were looking around for larger facilities.

In the meantime a new Baptist group was organized under the leadership of the Reverend J. Mahlon Kirkland, who had completed an eighteen-year pastorate at the First Baptist Church in Georgetown. Interestingly enough, this new group held its initial meeting on December 31, 1978, in the parish house of Prince George Church, a meeting site offered by Prince George's new rector, the Reverend Melvin R. Hyman. Naming itself the New Dawn Mission, this group began on March 4, 1979, to hold services in this structure, once the Church of the Messiah, at the invitation of the Reverend David B. Coker, pastor of the Maryville Pentecostal Holiness Church. The Baptists took a liking to the building and purchased it just before the Pentecostal congregation moved to its new church in late July, 1979. On July 23, 1979, the Baptists moved the building from Poplar Street to the present site, the corner of Old Charleston Road and Maple Street in Maryville. The congregation on June 27, 1979, had voted to change its designation officially from New Dawn Mission to Church of the Messiah. The building was opened once again for services on August 29, 1979.

Since then the adjacent educational building has been constructed, and the congregation has grown significantly. The Church of the Messiah is now affiliated with the Southern Baptist Convention. It is still under the leadership of the Reverend J. Mahlon Kirkland and has every reason to look forward to an ever-expanding future. The building has been fully restored. The congregation is aware of the structure's historic heritage and is taking care to preserve it in an appreciative spirit.

Hide Out
(Wicklow Hall)

In recent years the house here has undergone some renovation, and the grounds have been cleared. The major changes, though, appear to be in ownership and name.

T. Cordes Lucas lived here with his family from 1927 when the place was deeded to him by the Kinloch Gun Club until 1962 when he moved to McClellanville. He died on May 9, 1971. With his death, the Santee lost one of its most delightful, most informed and most irreplaceable authorities on local history and rice planting.

This plantation changed hands before Mr. Lucas's death, however. The three Lucas daughters, Alice Lucas Hutson, Mary Lucas Easley and Katherine Lucas Bull, who held title to the property, sold the nearly nine-acre tract in a transaction finally completed on July 31, 1968, to Mr. and Mrs. Maurice Thatcher Paris of Columbia, South

Carolina. The Parises made the plantation their home. They also reverted to the use of an earlier name once attached to this property and officially redesignated it Wicklow Hall. On December 31, 1976, the Parises deeded the place to their daughter Katrina and her husband Eugene S.N. Lawrimore, a Georgetown attorney. The Lawrimores then made Wicklow Hall their year-round residence. On December 21, 1988, the Lawrimores sold the place to present owner Nancy duPont Reynolds, one of the duPont heirs to adjacent Kinloch plantation. A resident of Greenville, Delaware, Mrs. Reynolds will use the property for winter vacations.

Because Wicklow Hall has been the subject of occasional bits of inaccurate information over the past decade or so, it might be well to attempt as much clarification as possible. First of all, the house was modernized in the 1930's by Mr. and Mrs. T. Cordes Lucas, not later, as some reports have indicated.

Additionally, an outbuilding thought by some to be an old schoolhouse was really a playhouse constructed for the oldest Lucas daughter, Alice, in the 1930's. Furthermore, Mr. Lucas utilized the whole house during his last years here, not just one room as one publication stated.

The Paris family, besides freshening up the house and clearing the grounds, also restored an outbuilding believed to be the old kitchen. The property has remained intact and largely unchanged for some time. From 1969 until 1971, Mr. and Mrs. Paris opened the place to the public.

Wicklow Hall was placed on the National Register of Historic Places in 1978.

Kinloch

The principal change at Kinloch plantation is in ownership. The very colorful Robert Edward (Ted) Turner, III, chairman of the board and president of the Atlanta-based Turner Broadcasting System, is the present owner. He purchased this large estate on December 28, 1982, from the heirs of Delaware industrialist Eugene E. duPont, who had owned the property for years with his cousin Eugene duPont. At the time of his death in 1966, Eugene E. was sole owner of Kinloch, having bought out his cousin's interest in 1962, and had bequeathed it to his four daughters, Deo duPont Weymouth, Phyllis duPont Schutt, Murton duPont Carpenter and Nancy duPont Reynolds, all of Greenville, Delaware. By the time of the sale to Ted Turner, Mrs. Weymouth had died so her heirs joined the other three duPont daughters in transferring the Kinloch estate to Mr. Turner.

Like the duPonts, Ted Turner uses the property primarily for hunting and wildlife conservation. Consisting of 9,000 acres that were once thirteen individual plantations, Kinloch has been since the days of the Kinloch Gun Club one of the choicest hunting areas on the Santee. However, Kinloch is more than a place for hunting. For years wildlife conservation has been practiced here. Modern game management techniques are used to provide suitable wetlands for wintering waterfowl. Moreover, the forests here are managed to promote the habitation of quail, turkey, doves and deer.

Ted Turner's television company, though headquartered in Atlanta, is national in scope. Among its operations is the Cable News Network. Mr. Turner himself is frequently in the news. Not only is he a successful, enterprising business executive, but he is also a noted sportsman and a recognized conservationist.

Members of the Turner family spend time each year at Kinloch, where the lodge and planted grounds continue to be attractively maintained.

Woodside

Woodside plantation is enjoying new life under present owners Mr. and Mrs. Herbert J. Butler of Charleston. Back in 1956, C. Porter Schutt, son-in-law of Eugene E. duPont of neighboring Kinloch plantation, purchased the property from Mr. and Mrs. Russell M. Doar, who retained lifetime use of the old house here. Following Mr. Doar's death in 1960, Mrs. Doar moved to Georgetown, leaving the dwelling unoccupied. This house—the original Woodside house, believed to have been constructed in 1843—was dismantled in 1970 because of its dilapidated condition. Finally in January of 1983, the Butlers purchased the plantation, consisting at the time of 155 acres, from C. Porter Schutt.

During the following year the new owners completed construction of the present Woodside house, an attractive dwelling that incorporates a typical Lowcountry plantation design of the rice era. The wing was added in 1987. Situated on a knoll overlooking old North Santee rice fields, the house offers an impressive view of the surrounding area. Mr. and Mrs. Butler spend considerable time all through the year at Woodside, happily a place of activity again.

Herbert Butler, a Charleston business executive, also owns other property on the Santee. At the present time he has a 400-acre tract and a 850-acre tract, both upriver from Woodside.

Rice Hope

Changes in size and ownership have been the major developments at Rice Hope since the long ownership of the late William N. Beach.

Mr. Beach purchased the property, consisting then of 1,113 acres in 1926 and through the years added various tracts. By the time of his death in 1955, he had enlarged Rice Hope's boundaries tremendously, absorbing thousands of acres from adjoining plantations.

Since 1955 the trend has reversed. Koppers Company, Incorporated, purchased a large chunk of timberland in 1956. Following Mrs. William N. Beach's death in 1960, the remaining portion was sold. Williams Furniture Corporation of Sumter, South Carolina, obtained this part, including the house and grounds, in the fall of 1961 from the Beach estate. Williams then split off the house and 237 surrounding acres and sold them in 1969 to a Charleston group composed of Thomas G. Buist, Herbert J. Butler, A.E. Geer, Jr., B.H. Rutledge Moore, C. Allen Smith and D. Van Smith. After using the plantation for hunting and weekend retreats for some thirteen years, the Charleston group sold this part of Rice Hope—that is, the house and the accompanying 237 acres—to the present owners, Rice Hope Partners, on June 9, 1982. Rice Hope Partners is a group consisting of the following men: Rutherford Galloway and William Lowrance, both of Easley, South Carolina; Paul Autry, III, Darrell Jervey, Graham Proffit, David Douglas and William McCauley, all of Greenville, South Carolina; and Randy Shelley of Florence, South Carolina.

The present owners use Rice Hope for the same purposes as the preceding owners—for hunting and visiting. The grounds here are still quite lovely and make the place as appealing as ever.

Hopsewee

Not only one of the oldest plantation houses in Georgetown County, Hopsewee is one of the most important as well. Constructed between 1735 and 1740, this fine, old dwelling was the birthplace in 1749 of Thomas Lynch, Jr., a South Carolina signer of the Declaration of Independence.

Today the house not only exists; it is in excellent condition. The grounds are cleared and planted, and the old outdoor kitchens are restored. These are open to the public through the graciousness of the present owners, Mr. and Mrs. James T. Maynard, who make this historic place their home. Fortunately for everyone, the Maynards' appreciation of history and sense of stewardship have motivated them to share Hopsewee with their fellow citizens.

This arrangement had its beginning in the purchase of the Hopsewee house and surrounding four and one-half acres by the Maynards, Georgetown residents, from Mrs. Reading Wilkinson in May of 1969. Mrs. Wilkinson and her husband the late Colonel Wilkinson had obtained this part of the plantation in 1947 from the International Paper Company of Georgetown. IPC retained the rest of the 781-acre tract. Mr. and Mrs. Maynard followed up their initial purchase by later getting from IPC fifty more acres adjacent to their property line. In the summer of 1988, they bought from IPC an additional thirty acres from the original Hopsewee tract. The total acreage around the house is now eighty-four and one-half acres.

Immediately after moving to Hopsewee at the end of May, 1969, Mr. and Mrs. Maynard began renovating the house. They officially opened it to the public in 1970 on a year-round basis.

Later, they repaired the two shingled structures directly back of the main house. Both of these are believed to be the old kitchens because each has a large fireplace on one side of the chimney and a double fireplace on the other side. The presence of two outdoor kitchens is unusual. Through the years the Maynards have located the chimney remains of twelve slave cabins along what was obviously the old street some one hundred yards to the rear and slightly west of the Hopsewee house. The slave street ran perpendicular to the river. After the Civil War, the timbers and bricks in the cabins gradually disappeared.

Hopsewee's handsome dwelling presents a superb example of early eighteenth century Lowcountry plantation architecture. Built of durable black cypress, it features remarkable balance in room arrangement. There are four rooms on each of the two floors, and each floor has a wide center hall extending from front to back, opening the whole house to the cool breeze from just about any direction. Most definitely, the elder Thomas Lynch incorporated his knowledge of the Santee to good effect in the construction of this house. It is comfortable by even today's standards. However, the house goes considerably beyond comfort. Each room is nicely enhanced by handcarved molding. Furthermore, the staircase is no ordinary staircase. It is Georgian in style and quite lovely. The old heart pine floors which the Maynards refinished are also outstanding.

Hopsewee was placed on the National Register of Historic Places in 1971. It was awarded even greater recognition in 1972 by the United States Department of the Interior, which designated the place that year as a Registered National Historic Landmark.

NOTES

Notes

WACCAMAW NECK

[1]Personal interview with Bernard M. Baruch at Little Hobcaw, Kingstree, S.C., January 22, 1954.

[2]Henry A. M. Smith, "Hobcaw Barony," *South Carolina Historical and Genealogical Magazine,* XIV (April, 1913), 61-62. For the sake of brevity, this magazine will be referred to as *SCH&GM.*

[3]*Ibid.,* p. 62.

[4]*Ibid.*

[5]*Ibid.,* p. 63.

[6]*Ibid.,* pp. 63-64.

[7]*Ibid.,* p. 64.

[8]*Ibid.,* p. 65.

[9]*Ibid.*

[10]*Ibid.,* pp. 65, 73. H. D. Bull, "Kinloch of South Carolina," *SCH&GM,* XLVI (July, 1945), 160.

[11]Smith, "Hobcaw Barony," *SCH&GM,* XIV, 65-66.

[12]*Ibid.,* p. 66.

[13]*Ibid.*

[14]*Ibid.*

[15]Deed Book E, p. 339, Courthouse, Georgetown, S. C.

[16]Letter from Malcolm R. Donaldson, son of Sidney T. Donaldson, March 24, 1954. Cross Index to Conveyances of Real Estate, Courthouse, Georgetown, S.C.

[17]Deed Book X, p. 230, Courthouse, Georgetown, S.C.

[18]Will of Colonel William Alston recorded in office of judge of probate, Charleston County Courthouse, Charleston, S.C.

[19]Copy of the will of Colonel Joshua John Ward and legal papers attached to this copy in the possession of Colonel Ward's grandson, Joshua John Ward, of Pawleys Island, S.C. Also Deed Book M, p. 101, Courthouse, Georgetown, S.C.

[20]Letter from Malcolm R. Donaldson, March 24, 1954.

[21]Letter from Bernard Baruch, January 20, 1954.

[22]*Ibid.*

[23]Carter Field, *Bernard Baruch: Park Bench Statesman* (New York: McGraw-Hill Book Co., Inc., 1944), p. 133.

[24]From *The Carolina Field* (May 24, 1905). A copy of this publication was shown the writer by Miss Aileen Donaldson, Georgetown, S.C. Details of the

Cleveland–Caines story were given by Mrs. Charles Weston Rosa of Georgetown. A prominent member of the Daughters of the American Revolution, Mrs. Rosa was an earnest student of county history for many years and was responsible for the placement of historical markers along the highways of Georgetown County.

[25]Arney R. Childs, ed., *Rice Planter and Sportsman: The Recollections of J. Motte Alston, 1821-1909* (Columbia: University of South Carolina Press, 1953), p. 37.

[26]For a fuller account of the land transactions see Hobcaw Barony.

[27]Letter from Bernard Baruch, January 20, 1954.

[28]Elizabeth Deas Allston, *Allstons and Alstons of Waccamaw* (copyrighted 1936), p. 8. Smith, "Hobcaw Barony," *SCH&GM*, XIV, 69.

[29]Allston, *Allstons and Alstons of Waccamaw*, pp. 8, 16-18.

[30]*Ibid.*, pp. 27, 35.

[31]Quoted *ibid.*, p. 84.

[32]Henry DeSaussure Bull, *All Saints Church, Waccamaw, 1739-1948* (Published by the Historical Activities Committee of the South Carolina Society of Colonial Dames of America, 1948), pp. 14, 18.

[33]Allston, *Allstons and Alstons of Waccamaw*, p. 35.

[34]Copy of the will of Colonel Joshua John Ward.

[35]Deed Book M, p. 87, Courthouse, Georgetown, S.C.

[36]*Ibid.*, p. 113.

[37]Deed Book A-1, p. 53, Courthouse, Georgetown, S.C.

[38]Deed Book M, p. 374, Courthouse, Georgetown, S.C.

[39]Allston, *Allstons and Alstons of Waccamaw*, p. 38.

[40]Deed Book M, p. 382, Courthouse, Georgetown, S.C.

[41]Allston, *Allstons and Alstons of Waccamaw*, p. 27.

[42]Susan Lowndes Allston, *Brookgreen, Waccamaw, in the Carolina Low Country* (copyrighted 1935), p. 17.

[43]Smith, "Hobcaw Barony," *SCH&GM*, XIV, 64, 69.

[44]Allston, *Allstons and Alstons of Waccamaw*, p. 22.

[45]*President Washington's Diaries 1791 to 1799*, transcribed and compiled by Jos. A. Hoskins (Summerfield, N.C., 1921), p. 25.

[46]Allston, *Allstons and Alstons of Waccamaw*, p. 25.

[47]Deed Book M, p. 384, Courthouse, Georgetown, S.C.

[48]Deed Book E-2, p. 120, Courthouse, Georgetown, S.C.

[49]Allston, *Allstons and Alstons of Waccamaw*, p. 91.

[50]*Ibid.*, pp. 47-91.

[51]Interview with Mrs. John E. Allston, Pawleys Island, S.C.

[52]Deed Book U-2, p. 44, Courthouse, Georgetown, S.C.

[53]See Waverly.

[54]Deed Book F-2, p. 328, Courthouse, Georgetown, S.C.

[55]Deed Book Y-2, p. 119, Courthouse, Georgetown, S.C.

[56]Captain William Ancrum, Pawleys Island, S.C.

[57]Deed Book J-3, p. 590, Courthouse, Georgetown, S.C.

[58]Deed Book Y-3, p. 228, Courthouse, Georgetown, S.C.

[59]A plat of Litchfield dated 1794 shows that plantation to be bordered south on lands of the "John Alston Est." Supporting the claim of the plat is a letter dated November 2, 1953, from Miss Susan Lowndes Allston, direct descendant of John Allston. A student of family history, Miss Allston writes that John Allston settled at Waverly in the early days. The Litchfield plat lists below the "John Alston Est." the name of "Capt. W. Alston." This cannot be Captain William Allston, John's son, as he died thirteen years before the plat was drawn; nor does his will mention lands that logically could be interpreted as Waverly. On the other hand, William Alston of Clifton, who was a captain in 1794 and who had already begun spelling his last name with a single *l*, definitely is known to have owned at least a large part of the plantation. A deed of 1871 found on page 400, Deed Book C, Courthouse, Georgetown, S.C., describes Waverly as then bordering north partly on lands of Colonel William Alston's estate. These same lands were shown on the 1794 plat of the adjacent Litchfield plantation as part of Waverly property, and over this entire area from river to sea as shown on that plat is written twice the name, "Capt. W. Alston." It follows, then, that all of Waverly must have been owned at one time by Colonel William Alston of Clifton. It is to be noted that the plat does not differentiate between the one-*l* and two-*l* Alstons.

[60]J. H. Easterby, *The South Carolina Rice Plantation: As Revealed in the Papers of Robert F. W. Allston* (Chicago: University of Chicago Press, 1945), pp. 13, 19.

[61]Elizabeth W. Allston Pringle, *Chronicles of Chicora Wood* (Boston: The Christopher Publishing House, 1940), p. 33; Easterby, *The South Carolina Rice Plantation,* p. 19.

[62]Easterby, *The South Carolina Rice Plantation,* pp. 25, 36.

[63]*Ibid.*, p. 6.

[64]*Ibid.*, pp. 6, 19.

[65]*Ibid.*, p. 36.

[66]Deed Book C, p. 400, Courthouse, Georgetown, S.C.

[67]Pringle, *Chronicles of Chicora Wood,* pp. 33-34.

[68]Deed Book C, p. 400.

[69]Henry A. Lachicotte, *Genealogy of the Rossignol–Lachicotte Family* (Douglas, Ga.: Julian W. Frier Publishing Co., 1950), p. 68.

[70]*Ibid.*, pp. 46-56.

[71]*Ibid.*, pp. 27, 40-50.

[72]This patent is in the possession of Arthur H. Lachicotte, Waverly plantation.

[73]See Rossdhu.

[74]Bull, *All Saints, Waccamaw,* pp. 7-8.

[75]*Ibid.*, p. 8.

[76]*Ibid.*, pp. 9-10.

[77]*Ibid.*, pp. 10-11.

[78]*Ibid.*, p. 16.

[79]*Ibid.*, pp. 16-17.

[80]*Ibid.*, pp. 18, 36.

[81]*Ibid.*, p. 19.

[82]*Ibid.*, pp. 21-22.

[83]*Ibid.*, p. 33.

[84]*Ibid.*, p. 37.

[85]*Ibid.*, p. 39.

[86]*Ibid.*, pp. 43-44.

[87]*Ibid.*, pp. 46-47, 51.

[88]*Ibid.*, pp. 51-52.

[89]*Ibid.*, p. 53.

[90]From a copy of the will of Daniel Tucker attached to papers entitled "John Tucker of Bermuda and South Carolina and his Descendants" by George Haig Tucker, Memphis, Tenn.

[91]William Ensign Lincoln, *Lincoln, Peace, Porter, Ayer and Related Families* (New York: The Kickerbocker Press, 1930), pp. 98-99.

[92]From a copy of the will of Daniel Tucker.

[93]Lincoln, *Lincoln, Peace, Porter, Ayer and Related Families,* pp. 100-1.

[94]Childs, *Rice Planter and Sportsman,* p. 34.

[95]"Memoirs of Frederick Adolphus Porcher," *SCH&GM,* XLII (January, 1946), 47-48.

[96]Lincoln, *Lincoln, Peace, Porter, Ayer and Related Families,* p. 108.

[97]Cross Index to Conveyances of Real Estate, Courthouse, Georgetown, S.C.

[98]Easterby, *The South Carolina Rice Plantation,* p. 158. For further information on Daniel Tucker and John Hyrne Tucker I, see Litchfield.

[99]Lincoln, *Lincoln, Peace, Porter, Ayer and Related Families,* p. 102.

[100]Deed Book E, p. 732, Courthouse, Georgetown, S.C.

[101]Deed Book I, p. 152, Courthouse, Georgetown, S.C.

[102]Deed Book K, pp. 505, 698, Courthouse, Georgetown, S.C.

[103]Deed Book E-2, p. 98, Courthouse, Georgetown, S.C.

[104]Deed Book X-1, p. 370, Deed Book O-2, p. 92, Deed Book R-2, p. 425, Courthouse, Georgetown, S.C.

[105]Deed Book D-3, p. 363, Deed Book K-3, p. 137, Deed Book Z-3, p. 175, Courthouse, Georgetown, S.C.

[106]Information provided by Miss Susan Lowndes Allston.

[107]Allston, *Brookgreen, Waccamaw,* pp. 8-9, 11, 17.

[108]*Ibid.*, pp. 13-14.

[109]Edgar P. Richardson, *Washington Allston: A Study of the Romantic Artist in America* (Chicago: University of Chicago Press, 1948), pp. 5-7.

[110]*Ibid.*, pp. 134-35, 179-80.

[111]Allston, *Brookgreen, Waccamaw,* pp. 17-18, 22-23.

[112]Will of Colonel Ward (see note 19, above); Allston, *Brookgreen, Waccamaw,* pp. 21-23.

[113]Allston, *Brookgreen, Waccamaw,* p. 28-31.

[114]Smith, "Hobcaw Barony," *SCH&GM,* XIV, 75.

[115]*Brookgreen Gardens History* (pamphlet prepared by Brookgreen Gardens, S.C., 1945).

[116]*Brookgreen Gardens Sculpture by Anna Hyatt Huntington* (pamphlet prepared by Brookgreen Gardens, S.C., 1937).

[117]Beatrice Gilman Proske, *Robert A. Baillie: Carver of Stone* (Brookgreen, S.C.: printed by order of the trustees, 1946), p. 17.

[118]Smith "Hobcaw Barony," *SCH&GM*, XIV, 74.

[119]The map of Georgetown District in Robert Mills' *Atlas of South Carolina*, published in 1825, lists the name of Belin at the spot occupied by the present Wachesaw plantation.

[120]Copy of the will of Rev. James L. Belin. Also page 5 of a copy of the South Carolina Methodist Conference Minutes, 1859, Greenville, S.C.

[121]Information provided by Mrs. Genevieve Willcox Chandler.

[122]Bull, *All Saints, Waccamaw*, p. 36.

[123]*Ibid.*, p. 41.

[124]Will of Colonel Joshua John Ward (see note 19, above).

[125]Bull, *All Saints, Waccamaw*, p. 91.

[126]Deed Book P, p. 678, Deed Book G, p. 111, Deed Book X, p. 113, Courthouse, Georgetown, S.C.

[127]Deed Book C-1, p. 319, Deed Book K-2, p. 168, Courthouse, Georgetown, S.C.

[128]Mrs. William A. Kimbel, Wachesaw plantation.

[129]Letter from Dr. Chapman J. Milling dated Aug. 2, 1954.

[130]Mrs. William A. Kimbel.

[131]Deed Book K-2, p. 167, Courthouse, Georgetown, S.C.

[132]An 1851 plat of Wachesaw, Plat Book B, p. 100, Courthouse, Georgetown, S.C.

BLACK RIVER

[1]*Abstract of the Title of Mrs. Mary A. L. Ford to Wedgefield*, prepared by Walter Hazard, Sept. 30, 1911. Considered a brilliant attorney, the late Walter Hazard of Georgetown did outstanding research in tracing titles of various plantations in Georgetown County. His work has been accepted through the years as being of the utmost reliability.

[2]James R. Parker, Georgetown, S.C., from his notes based on a study of papers concerned with Windsor. This statement is supported also by Harvey Toliver Cook in his book, *Rambles in the Pee Dee Basin South Carolina* (Columbia, S.C.: The State Co., 1926), p. 44.

[3]Walter Hazard, *Abstract of Title to Wedgefield*.

[4]James R. Parker, Georgetown, S.C.

[5]*Ibid.*

[6]*Ibid.*

[7]Deed Book C, p. 100, Deed Book I, p. 269, Courthouse, Georgetown, S.C.
[8]Easterby, *The South Carolina Rice Plantation,* p. 18.
[9]Pringle, *Chronicles of Chicora Wood,* p. 249.
[10]Deed Book H-2, p. 274, Courthouse, Georgetown, S.C.
[11]*The State* (Columbia, S.C.), July 12, 1954.
[12]Mrs. Paul D. Mills, Windsor plantation.
[13]Letter from Miss Susan Lowndes Allston dated Nov. 2, 1953.
[14]Mrs. Paul D. Mills, Windsor plantation.
[15]See Windsor.
[16]Walter Hazard, *Abstract of Title to Wedgefield.*
[17]James R. Parker, Georgetown, S.C., from notes based on a study of legal and family papers concerned with Wedgefield.
[18]Walter Hazard, *Abstract of Title to Wedgefield.*
[19]Deed Book S-1, p. 300, Courthouse, Georgetown, S. C.
[20]Deed Book Y-1, p. 25, Courthouse, Georgetown, S.C.
[21]Deed Book B-2, p. 103, Courthouse, Georgetown, S.C.
[22]Deed Book O-2, p. 277, Courthouse, Georgetown, S.C.
[23]Deed Book R-2, p. 76, Courthouse, Georgetown, S.C.
[24]Mrs. Frederick Wentworth Ford, Georgetown, S.C.
[25]Deed Book J-3, p. 147, Courthouse, Georgetown, S.C.
[26]Deed Book P-3, p. 157, Courthouse, Georgetown, S.C.
[27]Deed Book X-3, p. 91, Courthouse, Georgetown, S.C.
[28]All data on Mansfield covering the years from 1718 to 1912 came from an article by James R. Parker, which appeared in *The Georgetown Times* on March 28, 1941. A grandson of Dr. Francis S. Parker, James R. Parker based his article on an extensive study of family papers, grant books in the Secretary of State's office, Columbia, S.C., will books in the Probate Judge's office, Charleston, S.C., and material found in library records.
[29]Deed Book E-1, p. 308,Courthouse, Georgetown, S.C.
[30]Deed Book L-2, p. 47, Courthouse, Georgetown, S.C.
[31]Miss Minnie T. Hazard, secretary for Mansfield plantation.
[32]*The News and Courier* (Charleston, S.C.), Jan. 23, 1949.
[33]Letter from Colonel Montgomery to Kenneth Anderson, Clerk of the House of Representatives, Committee on Military Affairs, Washington, D.C., April 18, 1938.
[34]Cook, *Rambles in the Pee Dee Basin South Carolina,* p. 43.
[35]Helen Kohn Hennig, *Great South Carolinians* (Chapel Hill: University of North Carolina Press, 1940), pp. 85-87.
[36]*Ibid.,* pp. 92-105.
[37]Deed Book A, p. 200, Courthouse, Georgetown, S.C.
[38]Robert Mills, *Atlas of the State of South Carolina* (published 1825).
[39]Deed Book A, p. 200, Deed Book E-1, p. 124, Courthouse, Georgetown, S.C.
[40]Deed Book M-1, p. 93, Deed Book L-2, p. 104, Deed Book Z-3, p. 13, Courthouse, Georgetown, S.C.

[41]Deed Book C-4, p. 101, Courthouse, Georgetown, S.C.

[42]Plat Book C, p. 38, Courthouse, Georgetown, S.C.

[43]Pringle, *Chronicles of Chicora Wood,* pp. 156-57.

[44]Letter dated Aug. 15, 1954, from Miss Susan Lowndes Allston, niece of Elizabeth Allston Pringle.

[45]Easterby, *The South Carolina Rice Plantation,* p. 205.

[46]Pringle, *Chronicles of Chicora Wood,* pp. 255-56.

[47]*Ibid.,* pp. 261-63.

[48]Deed Book H, p. 393, Courthouse, Georgetown, S.C.

[49]*The News and Courier* (Charleston, S.C.), Sept. 15, 1954.

[50]Deed Book I, p. 162, Deed Book M-1, p. 480, Deed Book R-2, p. 35, Courthouse, Georgetown, S.C.

[51]*The State* (Columbia, S.C.), Sept. 21, 1954.

[52]Mrs. G. T. Skinner, Georgetown, S.C.

[53]*House and Garden Magazine,* February, 1938, pp. 34-35.

[54]Deed Book C, pp. 173-75, Courthouse, Georgetown, S.C.

[55]*Ibid.,* p. 51.

[56]*Ibid.*

[57]Deed Book Q, p. 418, Deed Book E-1, p. 567, Courthouse, Georgetown, S.C.

[58]Nat Kaminski, Jr. of Georgetown, S.C., grandson of Mrs. Heiman Kaminski.

[59]Deed Book N-3, p. 115, Deed Book 11, p. 489, Courthouse, Georgetown, S.C.

PEEDEE RIVER

[1]Pringle, *Chronicles of Chicora Wood,* p. 210.

[2]Deed Book C, pp. 173-75, Courthouse, Georgetown, S.C.

[3]*Ibid.* The writer is indebted to Dr. J. H. Easterby, Director of the South Carolina Archives Department for calling her attention to certain information in this deed.

[4]Mills, *Atlas of South Carolina.*

[5]Easterby, *The South Carolina Rice Plantation,* p. 21.

[6]Pringle, *Chronicles of Chicora Wood,* pp. 209-10.

[7]*Ibid.,* pp. 264-67.

[8]Deed Book C, p. 173.

[9]Deed Book L, pp. 471-76, Courthouse, Georgetown, S.C.

[10]Deed Book N-1, p. 36, Courthouse, Georgetown, S.C.

[11]Plat Book D, p. 53, Courthouse, Georgetown, S.C.

[12]Deed Book J-3, p. 507.

[13]Langdon Cheves, "Izard of South Carolina," *SCH&GM,* II (1901), 236-40.

[14]Deed Book F, p. 448, Courthouse, Georgetown, S.C.

[15]Deed Book U, p. 12, Deed Book D-1, p. 209, Deed Book C-3, p. 363, Courthouse, Georgetown, S.C.

[16]Deed Book C-4, p. 102, Deed Book B-4, p. 467, Courthouse, Georgetown, S.C.

[17]The late Mrs. Charles Weston Rosa (the former Miss Lila Sparkman) of Georgetown. Mrs. Rosa had in her home an oil painting of Dirleton Castle ruins.

[18]Deed Book N-1, p. 184, Courthouse, Georgetown, S.C., also records of Miss Mary A. Sparkman, Charleston, S.C.

[19]Easterby, *The South Carolina Rice Plantation,* pp. 203, 348.

[20]Mrs. Charles Weston Rosa.

[21]Thomas G. Samworth, Dirleton plantation.

[22]Easterby, *The South Carolina Rice Plantation,* pp. 211, 225.

[23]*Ibid.,* pp. 345-50.

[24]Minutes of the Manigault Light Dragoons now in the possession of Misses Helen and Julia Rosa, Georgetown, S.C.

[25]Deed Book E-1, p. 460, Courthouse, Georgetown, S.C.

[26]Deed Book A-2, p. 42, Courthouse, Georgetown, S.C.

[27]Deed Book L-2, p. 295, Courthouse, Georgetown, S.C.

[28]Deed Book F-3, p. 213, Courthouse, Georgetown, S.C.

[29]Deed Book I-3, p. 355, Courthouse, Georgetown, S.C.

[30]Mrs. Herman Carraway, Arundel plantation.

[31]Conveyances of Arundel from the heirs of Frederick Shaffer down through the property's purchase by Guendalos Rice Company are mentioned in the deed of sale from Louis Claude Lachicotte to Guendalos Rice Company. This deed was in the possession of Mrs. Herman Carraway at Arundel.

[32]Bull, *All Saints, Waccamaw,* p. 85.

[33]See Dirleton.

[34]Bull, *All Saints, Waccamaw,* pp. 91, 92.

[35]"Text of Bi-centennial Address by Bishop Albert S. Thomas at Prince Frederick's Church Last Sunday," *The Georgetown Times,* April 13, 1934. Bishop Thomas, is historiographer of the Episcopal Diocese of South Carolina.

[36]Chapman J. Milling, *Exile Without An End* (Columbia, S.C.: Bostick and Thornley, Inc., 1943), pp. 40-41.

[37]All above data used in this sketch with the exception of the paragraph on the Acadians was taken from Bishop Thomas' "Bi-centennial Address." See note 35, above.

[38]Easterby, *The South Carolina Rice Plantation,* pp. 19-20.

[39]*Ibid.,* pp. 10-20.

[40]Miss Susan Lowndes Allston, granddaughter of R. F. W. Allston. Miss Allston emphasizes, however, the uncertainty of this information.

[41]Easterby, *The South Carolina Rice Plantation,* pp. 19-20.

[42]Miss Susan Lowndes Allston.

[43]Easterby, *The South Carolina Rice Plantation,* pp. 12-13, 20-21.

[44]*Ibid.,* p. 12.

[45]Pringle, *Chronicles of Chicora Wood,* pp. 4-5.

[46]*Ibid.*, p. 6.

[47]Easterby, *The South Carolina Rice Plantation,* p. 18.

[48]Pringle, *Chronicles of Chicora Wood,* pp. 12, 35-36.

[49]Easterby, *The South Carolina Rice Plantation,* pp. 13-15.

[50]Pringle, *Chronicles of Chicora Wood,* p. 24.

[51]*Ibid.*, pp. 209-10.

[52]Easterby, *The South Carolina Rice Plantation,* pp. 21-22.

[53]Pringle, *Chronicles of Chicora Wood,* pp. 209-10.

[54]Easterby, *The South Carolina Rice Plantation,* p. 23.

[55]Pringle, *Chronicles of Chicora Wood,* pp. 268-69.

[56]Easterby, *The South Carolina Rice Plantation,* p. 18.

[57]New York: The Macmillan Co., 1913.

[58]New York: Charles Scribner's Sons, 1922.

[59]Deed Book E-2, p. 249, Courthouse, Georgetown, S.C.

[60]Mrs. D. C. Waddell, Chicora Wood plantation.

[61]Mills, *Atlas of South Carolina.*

[62]Mrs. H. D. Oswald, Georgetown, S.C.

[63]Mrs. Herman Carraway, Arundel plantation. Mrs. Carraway is also a descendant of Davison McDowell.

[64]Easterby, *The South Carolina Rice Plantation,* p. 52.

[65]From records in the possession of Miss Emma B. Richardson, of Charleston, S.C., descendant of Davison McDowell.

[66]From a memorandum dated April 26, 1906, prepared by the late Walter Hazard for J. Louis LaBruce on deeds and papers relating to LaBruce plantations. This memorandum is among other legal documents at Arundel plantation. The writer is indebted also to Mr. J. H. Easterby, Director of the South Carolina Archives Department, for certain portions of this information.

[67]See Chicora Wood.

[68]Easterby, *The South Carolina Rice Plantation,* pp. 18, 23.

[69]Deed Book I-3, p. 354, Courthouse, Georgetown, S.C.

[70]Mrs. Thomas S. Ragsdale, Exchange plantation.

[71]Elizabeth Allston Pringle, *A Woman Rice Planter* (New York: The Macmillan Co., 1914), pp. 53, 68.

[72]Cook, *Rambles in the Pee Dee Basin,* p. 80.

[73]Mills, *Atlas of South Carolina.*

[74]Mrs. Herman Carraway.

[75]Lincoln, *Lincoln, Peace, Porter, Ayer and Related Families,* p. 100.

[76]Pringle, *Chronicles of Chicora Wood,* pp. 158-59.

[77]Pringle, *A Woman Rice Planter,* pp. 68-69.

[78]Lincoln, *Lincoln, Peace, Porter, Ayer and Related Families,* p. 100.

[79]Deed Book O-1, p. 255, Courthouse, Georgetown, S.C.

[80]Cross Index to Conveyances of Real Estate, Courthouse, Georgetown, S.C.

[81]Deed Book J-3, p. 287, Courthouse, Georgetown, S.C.

Notes

SAMPIT RIVER AND WINYAH BAY

[1]Mrs. Charles Weston Rosa, "Friendfield on the Sampit," a paper in Georgetown County Memorial Library, Georgetown, S.C. The maiden name of Francis Withers' mother was obtained from Vol. XXXI, *SCH&GM*, by Miss Emma B. Richardson, a descendant of the Withers family connection. Mr. P. C. McClary, Jr., manager of the plantation, also provided certain information.

[2]Mrs. Charles Weston Rosa, "Northampton," a paper in Georgetown County Memorial Library, Georgetown, S.C.

[3]This account of the Withers tragedy on North Island in the hurricane of 1822 is taken from *The Charleston Courier,* Sept. 30, Oct. 3, 4, and 8, 1822.

[4]Mrs. Rosa, "Friendfield on the Sampit."

[5]Deed Book Q, p. 402, Courthouse, Georgetown, S.C.

[6]Deed Book R-1, p. 315, Courthouse, Georgetown, S.C.

[7]P. C. McClary, Jr.

[8]Deed Book K-2, p. 191, Courthouse, Georgetown, S.C.

[9]P. C. McClary, Jr.

[10]Mrs. Rosa, "Friendfield on the Sampit."

[11]Most of the data on Maryville plantation for the period of Read ownership came from Robert V. Read, Columbia, S.C. Mr. Read gave as his sources of information the wills of Francis Withers and John Harleston Read I, which are recorded in the judge of probate's office, Charleston County Courthouse, Charleston, S.C.; family and public records; Dr. Bull's *All Saints, Waccamaw*; and records of the South Carolina Historical Society, Miss Emma B. Richardson supplied certain clarifying information on the kinship between Mary Withers and James Withers. Page 453, Deed Book A, Courthouse, Georgetown, S.C., also helps to clarify this family connection. Information on Francis Withers' disposition of Harmony plantation was obtained from his will by Miss Richardson.

[12]Deed Book W-2, p. 739, Courthouse, Georgetown, S.C.

[13]Robert V. Read, son of John Harleston Read III.

[14]Deed Book W-2, p. 739.

[15]Deed Book H-3, p. 72, Courthouse, Georgetown, S.C.

[16]Henry A. M. Smith, "Winyah Barony," *SCH&GM,* XIII (Jan., 1912), 5-11. For a more detailed account of Winyah Barony, see Dover.

[17]A. S. Salley, ed., "Journal of General Peter Horry," *SCH&GM,* XXXVIII (April 1937), 49-50.

[18]Deed Book O, pp. 645-47, Courthouse, Georgetown, S.C.

[19]Data compiled by Miss Ellen Parker, Charleston, S.C. Miss Parker is a descendant of the Horry family.

[20]Salley, "Journal of General Peter Horry," *SCH&GM,* XXXVIII, 52.

[21]*Ibid.,* pp. 50-51.

[22]General Horry's will is on file in the judge of probate's office, Richland County Courthouse, Columbia, S.C.

[23]Deed Book O, pp. 645-47.

[24]Mabel L. Webber, "The Mayrant Family," *SCH&GM,* XXVII (April, 1926), 86-89.

[25]Deed Book B, p. 355, Courthouse, Georgetown, S.C.

[26]Webber, "The Mayrant Family," *SCH&GM,* XXXVIII, 89.

[27]Deed Book O, p. 645.

[28]T. Cordes Lucas, Hide Out plantation.

[29]Miss Margaret B. Rouquie, president of the Arthur Manigault chapter, United Daughters of the Confederacy. Miss Rouquie has been a student for many years of Confederate War history.

[30]Deed Book M-1, p. 549, Courthouse, Georgetown, S.C.

[31]Deed Book G-2, p. 218, Courthouse, Georgetown, S.C.

[32]Mrs. Henry M. Sage, Dover plantation.

[33]*Ibid.*

[34]Article by Francis Marion Kirk in *The News and Courier* (Charleston, S.C.), Dec. 22, 1935.

[35]Salley, "Journal of General Peter Horry," *SCH&GM,* XXXVIII (April, 1937), 49-50, (July, 1937), 82.

[36]Henry A. M. Smith, "Winyah Barony," *SCH&GM,* XIII, 5-8.

[37]*Ibid.,* pp. 9-11.

[38]*Ibid.,* p. 11.

[39]Salley, "Journal of General Peter Horry," *SCH&GM,* XXXVIII, 49-50. For a more detailed discussion of General Horry, see Belle Isle.

[40]Deed Book O, pp. 645-47, Courthouse, Georgetown, S.C.

[41]Will of General Peter Horry.

[42]Deed Book O, pp. 645-47.

[43]Webber, "The Mayrant Family," *SCH&GM,* XXVII, 86-89.

[44]Deed Book K, p. 161, Courthouse, Georgetown, S.C.

[45]Deed Book Y, p. 212, Courthouse, Georgetown, S.C.

[46]Webber, "The Mayrant Family," *SCH&GM,* XXVII, 89.

[47]Deed Book M-1, p. 552, Courthouse, Georgetown, S.C.

[48]*Ibid.*

[49]Mrs. Henry M. Sage.

[50]Deed Book M-1, p. 552.

[51]Mrs. Henry M. Sage.

[52]Webber, "The Mayrant Family," *SCH&GM,* XXVII, 89.

[53]Deed Book K, p. 161.

[54]Deed Book M, p. 9, Courthouse, Georgetown, S.C.

[55]Deed Book Y, p. 212.

[56]Deed Book M-1, p. 549, Courthouse, Georgetown, S.C.

[57]Deed Book R-2, p. 59, Courthouse, Georgetown, S.C.

[58]Mrs. Henry M. Sage. See Belle Isle for details on the Mendenhall house.

[59]*Ibid.*

[60]*The News and Courier,* Dec. 22, 1935.

[61]Mrs. Henry M. Sage.

[62]Deed Book E-3, p. 525, Courthouse, Georgetown, S.C.

[63]See Dover.

[64]Francis Johnstone, Athens, Ga., from family papers.

[65]Herbert Ravenel Sass, *Outspoken: 150 Years of the News and Courier* (Columbia: University of South Carolina Press, 1953), p. 9.

[66]Ulrich B. Phillips, *Life and Labor in the Old South* (Boston: Little, Brown and Co., 1929), p. 116. The writer is indebted to Mr. Sass for reference to this work containing a discussion of the tidal system of rice irrigation.

[67]Francis Johnstone.

[68]D. Huger Bacot, "Jonathan Lucas and Rice Culture," *South Carolina Magazine,* XVII (Nov., 1953), 11.

[69]Francis Johnstone.

[70]See Friendfield.

[71]Deed Book D, p. 417, Courthouse, Georgetown, S.C.; also Francis Johnstone.

[72]Francis Johnstone.

[73]Deed Book F, p. 173, Courthouse, Georgetown, S.C.

[74]Deed Book V, p. 271, Deed Book M-1, p. 353, Courthouse, Georgetown, S.C.

[75]T. Cordes Lucas.

NORTH SANTEE

[1]Map accompanying Henry A. M. Smith's history of Winyah Barony in the *SCH&GM,* XIII, 3-12.

[2]Cook, *Rambles in the Pee Dee Basin,* pp. 45-46.

[3]Deed Book M-1, p. 530, Courthouse, Georgetown, S.C.

[4]Miss Susan Lowndes Allston, granddaughter of Richard Henry Lowndes, from family records.

[5]*Ibid.* Also George B. Chase, *Lowndes of South Carolina* (Boston: A. Williams and Co., 1876), p. 23.

[6]Deed Book M-1, p. 530.

[7]Richard Lowndes Allston, grandson of Richard Henry Lowndes, Pawleys Island, S.C.

[8]Deed Book M-1, p. 530.

[9]Miss Susan Lowndes Allston. Also Chase, *Lowndes of South Carolina,* p. 41.

[10]Richard Lowndes Allston.

[11]Miss Susan Lowndes Allston.

[12]Deed Book M-1, p. 530.

[13]Deed Book M-1, p. 596, Courthouse, Georgetown, S.C.

[14]Deed Book D-2, p. 389, Courthouse, Georgetown, S.C.

[15]See Cat Island.

[16]Deed Book M-1, p. 530.

[17]Deed Book M-1, p. 596.

[18]Deed Book D-2, p. 389.

[19]T. Cordes Lucas.

[20]Direct Index to Conveyances of Real Estate, Courthouse, Georgetown, S.C.

[21]T. Cordes Lucas.

[22]Mrs. Horace B. Butterfield of Tacoma, Wash., granddaughter of Gen. E. P. Alexander.

[23]T. Cordes Lucas.

[24]Bacot, "Jonathan Lucas and Rice Culture," *South Carolina Magazine*, XVII, 11. Also T. Cordes Lucas and Francis Johnstone.

[25]See Estherville.

[26]Bacot, "Jonathan Lucas and Rice Culture," *South Carolina Magazine*, XVII, p. 11.

[27]See Estherville.

[28]See Friendfield.

[29]Francis Johnstone and T. Cordes Lucas.

[30]T. Cordes Lucas and C. E. Graham Reeves.

[31]Temporary Deed Book A, p. 96, Courthouse, Georgetown, S.C.

[32]Francis Johnstone.

[33]Henry Schulz Holmes, "The Trenholm Family," *SCH&GM*, XVI (1915), 153-54.

[34]Deed Book A, p. 202, Courthouse, Georgetown, S.C.

[35]Holmes, "The Trenholm Family," *SCH&GM*, XVI, 53.

[36]Deed Book Q, p. 446, Deed Book E-1, p. 100, Courthouse, Georgetown, S.C.

[37]Deed Book M-1, p. 552, Deed Book D-2, p. 84, Courthouse, Georgetown, S.C.

[38]Letter from C. E. Graham Reeves, Oct. 6, 1955.

[39]Deed Book R, p. 660, Courthouse, Georgetown, S.C.

[40]T. Cordes Lucas.

[41]Deed Book R, p. 660.

[42]Deed Book R-1, p. 304, Courthouse, Georgetown, S.C.

[43]T. Cordes Lucas. Miss Minnie Barnwell of Charleston, a descendant of Thomas Lynch, says that Lynchfield, Hopsewee, and Lynch Island, North Santee places, were formerly Thomas Lynch properties. This information is contained in a paper on Lynch plantations compiled by Miss Ellen Parker, a descendant of the Horry and Lynch families.

[44]Sass, *Outspoken*, pp. 63-64.

[45]Deed Book R-1, p. 304.

[46]T. Cordes Lucas.

[47]*Ibid.*

[48]Letter from the Rev. H. D. Bull, D.D., former rector of Prince George, Winyah, June 10, 1954.

[49]Samuel Gaillard Stoney, *Building a Church on the Santee 1804-1807* (Charleston, S.C.: The Charleston Museum, 1945), p. 4.

[50]Letter from the Rt. Rev. Albert S. Thomas, retired Bishop of the Diocese of South Carolina and presently diocesan historiographer, June 19, 1954.

[51]Stoney, *Building a Church,* p. 17.

[52]The Rt. Rev. Albert S. Thomas.

[53]Mrs. Frederick Wentworth Ford, who lived for many years on the North Santee. Bishop Thomas gives 1855 as the year that the present church was built, but this date probably was the year the old church was moved from North Santee ferry to the new site and re-erected. Mrs. Ford participated in the raising of funds at the turn of the century for construction of the church now standing.

[54]The Rt. Rev. Albert S. Thomas.

[55]Bull, *All Saints, Waccamaw,* pp. 27, 45.

[56]The Rev. H. D. Bull, from a diocesan journal.

[57]Mrs. Frederick Wentworth Ford.

[58]The Rt. Rev. Albert S. Thomas.

[59]The Rev. H. D. Bull.

[60]T. Cordes Lucas.

[61]Deed Book R, p. 788, Courthouse, Georgetown, S.C.

[62]T. Cordes Lucas.

[63]Deed Book R, p. 788.

[64]Deed Book E-1, pp. 291-94, Courthouse, Georgetown, S.C.

[65]T. Cordes Lucas.

[66]*The News and Courier,* May 9, 1954.

[67]T. Cordes Lucas.

[68]*Ibid.*

[69]A paper on Horry plantations compiled by Miss Ellen Parker.

[70]T. Cordes Lucas.

[71]Miss Ellen Parker.

[72]Deed Book E-1, pp. 291-94, Deed Book N, p. 508, Courthouse, Georgetown, S.C.

[73]Deed Book E-1, pp. 291-94.

[74]Deed Book N, p. 508.

[75]Deed Book E-1, pp. 291-94.

[76]T. Cordes Lucas.

[77]Deed Book G-2, p. 107, Courthouse, Georgetown, S.C.

[78]Deed Book E-1, pp. 291-94.

[79]*Ibid.*

[80]See Hide Out.

[81]T. Cordes Lucas.

[82]Russell Middleton Doar, Woodside plantation.

[83]Deed Book F, p. 239, Courthouse, Georgetown, S.C.

[84]*Ibid.*

[85]Mrs. Frederick Wentworth Ford.

[86]Mrs. Russell Middleton Doar.

[87]T. Cordes Lucas.

[88]Deed Book F, p. 239.

[89]T. Cordes Lucas.

[90]Bacot, "Jonathan Lucas and Rice Culture," *South Carolina Magazine,* XVII, 11.

[91]T. Cordes Lucas.

[92]From "Rice Hope," a paper by Mrs. Frederick Wentworth Ford, the former Mary Ashe Lucas. Mrs. Ford based her paper on family records.

[93]Bull, "Kinloch of South Carolina," *SCH&GM,* XLVI, 64-65.

[94]*Ibid.,* pp. 65-68.

[95]Ford, "Rice Hope."

[96]Bacot, "Jonathan Lucas and Rice Culture," *South Carolina Magazine,* XVII, 11.

[97]*Ibid.*

[98]Ford, "Rice Hope."

[99]Bacot, "Jonathan Lucas and Rice Culture," *South Carolina Magazine,* XVII, p. 11.

[100]Ford, "Rice Hope."

[101]*Ibid.*

[102]Deed Book E-2, p. 244, Courthouse, Georgetown, S.C.

[103]Article on Rice Hope by John M. Lofton, Jr., in *The News and Courier.*

[104]Ford, "Rice Hope."

[105]Mrs. Frederick Wentworth Ford.

[106]Deed Book G-2, p. 107.

[107]T. Cordes Lucas.

[108]Deed Book R-2, p. 59, Courthouse, Georgetown, S.C.

[109]Smith, "Winyah Barony," *SCH&GM,* XIII, 12.

[110]T. Cordes Lucas.

[111]"Thomas Lynch, Jr.," *Encyclopedia Americana* (1947), XVII, 741.

[112]Mrs. Reading Wilkinson, Hopsewee plantation, from data on Thomas Lynch, Jr.

[113]From "Hopsewee," a paper by the late Mrs. William Lucas, based on family records. This paper is in Georgetown County Memorial Library.

[114]*Encyclopedia Americana,* XVII, 741.

[115]Lucas, "Hopsewee."

[116]*Encyclopedia Americana,* XVII, 741.

[117]Lucas, "Hopsewee."

[118]T. Cordes Lucas.

[119]Lucas, "Hopsewee."

[120]*Ibid.*

[121]Deed Book H-3, p. 473, Deed Book K-3, p. 17, Courthouse, Georgetown. S.C.

[122]Deed Book N-3, p. 473, Courthouse, Georgetown, S.C.

[123]Lucas, "Hopsewee."

Index

Names of plantations appear in capitals. Married women are listed under their maiden names, when known.

Index

Index

Index

Index

Index

Index

Index

Supplementary Index

(Second through Seventh Printing)

Supplementary Index

Supplementary Index